Understanding Environmental Pollution

Understanding Environmental Pollution

Edited by
Diego Wells

Larsen & Keller
www.larsen-keller.com

Understanding Environmental Pollution
Edited by Diego Wells
ISBN: 978-1-63549-227-9 (Hardback)

☰ Larsen & Keller

Published by Larsen and Keller Education,
5 Penn Plaza,
19th Floor,
New York, NY 10001, USA

Cataloging-in-Publication Data

Understanding environmental pollution / edited by Diego Wells.
    p. cm.
Includes bibliographical references and index.
ISBN 978-1-63549-227-9
1. Pollution. 2. Pollution--Environmental aspects. 3. Pollution control equipment.
4. Pollution Prevention--Technological innovations. I. Wells, Diego.
TD174 .U53 2017
628.5--dc23

The publisher's policy is to use permanent paper from mills that operate a sustainable forestry policy. Furthermore, the publisher ensures that the text paper and cover boards used have met acceptable environmental accreditation standards.

Printed and bound in the United States of America.

For more information regarding Larsen and Keller Education and its products, please visit the publisher's website www.larsen-keller.com

Table of Contents

Preface

One of the major problems that our planet faces today is of pollution. It causes severe damage to natural resources and living organisms inhabiting the ecosystem. Pollution refers to the emission of harmful components like chemicals, methane gas, carbon dioxide, carbon monoxide, paint, grease and waste materials into environmental resources like water bodies, air, land, etc. This book is compiled in such a manner that it will provide in-depth knowledge about the causes and effects of pollution. The topics included in it are of utmost significance and are bound to provide incredible insights to readers. It explains all the important aspects of environmental pollution in the present day scenario. Those in search of information to further their knowledge will be greatly assisted by this textbook.

To facilitate a deeper understanding of the contents of this book a short introduction of every chapter is written below:

Chapter 1- Pollution is one of the symptoms of the rapid progress and urbanization that is being followed in most parts of the world. Pollution levels in the environment have reached dangerous levels and affects both living and non-living things. This chapter will provide an integrated understanding of pollution.

Chapter 2- Pollution can be broadly classified into a few categories. This chapter looks at air pollution, water pollution, soil contamination and noise pollution. Classification also leads to better management and the development of environment-friendly technology that can put a cap on pollution.

Chapter 3- Contamination of nature through human activity is the most dangerous threat to our natural resources. A lack of scientific and systematic pollution management leads to larger number of areas becoming polluted by contamination and waste accumulation. The effects of pollution are best understood in confluence with the major topics listed in the following chapter.

Chapter 4- As time passes and pollution levels remain unchecked, vulnerable areas of the Earth suffer from such accumulation of waste and the disruption of the natural cycles of the ecosystem. The ozone layer, the Arctic and increasing intensity of natural calamities are some of the aspects of rampant pollution. This chapter discusses the challenges of pollution in a critical manner providing key analysis to the subject matter.

Chapter 5- Environment-friendly technology encourages the use of long-lasting devices and cells for energy storage and consumption. It also focuses on re-use rather than throwing away otherwise non-biodegradable material. Tools and techniques are an important component of any field of study. The following chapter elucidates the various tools and techniques that are related to pollution control.

Finally, I would like to thank the entire team involved in the inception of this book for their valuable time and contribution. This book would not have been possible without their efforts. I would also like to thank my friends and family for their constant support.

Editor

Introduction to Pollution

Pollution is one of the symptoms of the rapid progress and urbanization that is being followed in most parts of the world. Pollution levels in the environment have reached dangerous levels and affects both living and non-living things. This chapter will provide an integrated understanding of pollution.

Pollution is the introduction of contaminants into the natural environment that cause adverse change. Pollution can take the form of chemical substances or energy, such as noise, heat or light. Pollutants, the components of pollution, can be either foreign substances/energies or naturally occurring contaminants. Pollution is often classed as point source or nonpoint source pollution.

The litter problem on the coast of Guyana, 2010

Ancient Cultures

Air pollution has always accompanied civilizations. Pollution started from prehistoric times when man created the first fires. According to a 1983 article in the journal *Science,* "soot found on ceilings of prehistoric caves provides ample evidence of the high levels of pollution that was associated with inadequate ventilation of open fires." Metal forging appears to be a key turning point in the creation of significant air pollution levels outside the home. Core samples of glaciers in Greenland indicate increases in pollution associated with Greek, Roman and Chinese metal production, but at that time the pollution was comparatively small and could be handled by nature.

Urban Pollution

The burning of coal and wood, and the presence of many horses in concentrated areas made the cities the cesspools of pollution. The Industrial Revolution brought an infusion of untreated chemicals and wastes into local streams that served as the water supply. King Edward I of England banned the burning of sea-coal by proclamation in London in 1272, after its smoke became a problem. But the fuel was so common in England that this earliest of names for it was acquired because it could be carted away from some shores by the wheelbarrow.

It was the industrial revolution that gave birth to environmental pollution as we know it today. London also recorded one of the earlier extreme cases of water quality problems with the Great Stink on the Thames of 1858, which led to construction of the London sewerage system soon afterward. Pollution issues escalated as population growth far exceeded view ability of neighborhoods to handle their waste problem. Reformers began to demand sewer systems, and clean water.

In 1870, the sanitary conditions in Berlin were among the worst in Europe. August Bebel recalled conditions before a modern sewer system was built in the late 1870s:

> "Waste-water from the houses collected in the gutters running alongside the curbs and emitted a truly fearsome smell. There were no public toilets in the streets or squares. Visitors, especially women, often became desperate when nature called. In the public buildings the sanitary facilities were unbelievably primitive....As a metropolis, Berlin did not emerge from a state of barbarism into civilization until after 1870."

The primitive conditions were intolerable for a world national capital, and the Imperial German government brought in its scientists, engineers and urban planners to not only solve the deficiencies but to forge Berlin as the world's model city. A British expert in 1906 concluded that Berlin represented "the most complete application of science, order and method of public life," adding "it is a marvel of civic administration, the most modern and most perfectly organized city that there is."

The emergence of great factories and consumption of immense quantities of coal gave rise to unprecedented air pollution and the large volume of industrial chemical discharges added to the growing load of untreated human waste. Chicago and Cincinnati were the first two American cities to enact laws ensuring cleaner air in 1881. Pollution became a major issue in the United States in the early twentieth century, as progressive reformers took issue with air pollution caused by coal burning, water pollution caused by bad sanitation, and street pollution caused by the 3 million horses who worked in American cities in 1900, generating large quantities of urine and manure. As historian Martin Melosi notes, The generation that first saw automobiles replacing the horses saw cars as "miracles of cleanliness.". By the 1940s, however, automobile-caused smog was a major issue in Los Angeles.

Other cities followed around the country until early in the 20th century, when the short lived Office of Air Pollution was created under the Department of the Interior. Extreme smog events were experienced by the cities of Los Angeles and Donora, Pennsylvania in the late 1940s, serving as another public reminder. Air pollution would continue to be a problem in England, especially later during the industrial revolution, and extending into the recent past with the Great Smog of 1952.

Awareness of atmospheric pollution spread widely after World War II, with fears triggered by reports of radioactive fallout from atomic warfare and testing. Then a non-nuclear event, The Great Smog of 1952 in London, killed at least 4000 people. This prompted some of the first major modern environmental legislation, The Clean Air Act of 1956.

Pollution began to draw major public attention in the United States between the mid-1950s and early 1970s, when Congress passed the Noise Control Act, the Clean Air Act, the Clean Water Act and the National Environmental Policy Act.

Smog Pollution in Taiwan

Severe incidents of pollution helped increase consciousness. PCB dumping in the Hudson River resulted in a ban by the EPA on consumption of its fish in 1974. Long-term dioxin contamination at Love Canal starting in 1947 became a national news story in 1978 and led to the Superfund legislation of 1980. The pollution of industrial land gave rise to the name brownfield, a term now common in city planning.

The development of nuclear science introduced radioactive contamination, which can remain lethally radioactive for hundreds of thousands of years. Lake Karachay, named by the World-watch Institute as the "most polluted spot" on earth, served as a disposal site for the Soviet Union throughout the 1950s and 1960s. Second place may go to the area of Chelyabinsk Russian as the "Most polluted place on the planet".

Nuclear weapons continued to be tested in the Cold War, especially in the earlier stages of their development. The toll on the worst-affected populations and the growth since then in understanding about the critical threat to human health posed by radioactivity has also been a prohibitive complication associated with nuclear power. Though extreme care is practiced in that industry, the potential for disaster suggested by incidents such as those at Three Mile Island and Chernobyl pose a lingering specter of public mistrust. Worldwide publicity has been intense on those disasters. Widespread support for test ban treaties has ended almost all nuclear testing in the atmosphere.

International catastrophes such as the wreck of the Amoco Cadiz oil tanker off the coast of Brittany in 1978 and the Bhopal disaster in 1984 have demonstrated the universality of such events and the scale on which efforts to address them needed to engage. The borderless nature of atmosphere and oceans inevitably resulted in the implication of pollution on a planetary level with the issue of global warming. Most recently the term persistent organic pollutant (POP) has come to describe a group of chemicals such as PBDEs and PFCs among others. Though their effects remain somewhat less well understood owing to a lack of experimental data, they have been detected in various ecological habitats far removed from industrial activity such as the Arctic, demonstrating diffusion and bioaccumulation after only a relatively brief period of widespread use.

A much more recently discovered problem is the Great Pacific Garbage Patch, a huge concentration of plastics, chemical sludge and other debris which has been collected into a large area of the Pacific Ocean by the North Pacific Gyre. This is a less well known pollution problem than the others described above, but nonetheless has multiple and serious consequences such as increasing wildlife mortality, the spread of invasive species and human ingestion of toxic chemicals. Organizations such as 5 Gyres have researched the pollution and, along with artists like Marina DeBris, are working toward publicizing the issue.

Pollution introduced by light at night is becoming a global problem, more severe in urban centres, but nonetheless contaminating also large territories, far away from towns.

Growing evidence of local and global pollution and an increasingly informed public over time have given rise to environmentalism and the environmental movement, which generally seek to limit human impact on the environment.

Forms of Pollution

The Lachine Canal in Montreal, Quebec, Canada.

The major forms of pollution are listed below along with the particular contaminant relevant to each of them:

- Air pollution: the release of chemicals and particulates into the atmosphere. Common gaseous pollutants include carbon monoxide, sulfur dioxide, chlorofluorocarbons (CFCs) and nitrogen oxides produced by industry and motor vehicles. Photochemical ozone and smog are created as nitrogen oxides and hydrocarbons react to sunlight. Particulate matter, or fine dust is characterized by their micrometre size PM_{10} to $PM_{2.5}$.

- Light pollution: includes light trespass, over-illumination and astronomical interference.

- Littering: the criminal throwing of inappropriate man-made objects, unremoved, onto public and private properties.

- Noise pollution: which encompasses roadway noise, aircraft noise, industrial noise as well as high-intensity sonar.

- Soil contamination occurs when chemicals are released by spill or underground leakage. Among the most significant soil contaminants are hydrocarbons, heavy metals, MTBE, herbicides, pesticides and chlorinated hydrocarbons.

- Radioactive contamination, resulting from 20th century activities in atomic physics, such as nuclear power generation and nuclear weapons research, manufacture and deployment.

- Thermal pollution, is a temperature change in natural water bodies caused by human influence, such as use of water as coolant in a power plant.

- Visual pollution, which can refer to the presence of overhead power lines, motorway billboards, scarred landforms (as from strip mining), open storage of trash, municipal solid waste or space debris.

Blue drain and yellow fish symbol used by the UK Environment Agency to raise awareness of the ecological impacts of contaminating surface drainage

- Water pollution, by the discharge of wastewater from commercial and industrial waste (intentionally or through spills) into surface waters; discharges of untreated domestic sewage, and chemical contaminants, such as chlorine, from treated sewage; release of waste and contaminants into surface runoff flowing to surface waters (including urban runoff and agricultural runoff, which may contain chemical fertilizers and pesticides); waste disposal and leaching into groundwater; eutrophication and littering.

- Plastic pollution: involves the accumulation of plastic products in the environment that adversely affects wildlife, wildlife habitat, or humans.

Pollutants

A pollutant is a waste material that pollutes air, water or soil. Three factors determine the severity of a pollutant: its chemical nature, the concentration and the persistence.

Cost of Pollution

Pollution has cost. Manufacturing activities that cause air pollution impose health and clean-up costs on the whole society, whereas the neighbors of an individual who chooses to fire-proof his home may benefit from a reduced risk of a fire spreading to their own houses. If external costs exist, such as pollution, the producer may choose to produce more of the product than would be produced if the producer were required to pay all associated environmental costs. Because responsibility or consequence for self-directed action lies partly outside the self, an element of externalization is involved. If there are external benefits, such as in public safety, less of the good may be produced than would be the case if the producer were to receive payment for the external benefits to others.

Sources and Causes

Air pollution comes from both natural and human-made (anthropogenic) sources. However, globally human-made pollutants from combustion, construction, mining, agriculture and warfare are increasingly significant in the air pollution equation.

Motor vehicle emissions are one of the leading causes of air pollution. China, United States, Russia, India Mexico, and Japan are the world leaders in air pollution emissions. Principal stationary pollution sources include chemical plants, coal-fired power plants, oil refineries, petrochemical plants, nuclear waste disposal activity, incinerators, large livestock farms (dairy cows, pigs, poultry, etc.), PVC factories, metals production factories, plastics factories, and other heavy industry. Agricultural air pollution comes from contemporary practices which include clear felling and burning of natural vegetation as well as spraying of pesticides and herbicides

About 400 million metric tons of hazardous wastes are generated each year. The United States alone produces about 250 million metric tons. Americans constitute less than 5% of the world's population, but produce roughly 25% of the world's CO_2, and generate approximately 30% of world's waste. In 2007, China has overtaken the United States as the world's biggest producer of CO_2, while still far behind based on per capita pollution - ranked 78th among the world's nations.

An industrial area, with a power plant, south of Yangzhou's downtown, China

In February 2007, a report by the Intergovernmental Panel on Climate Change (IPCC), representing the work of 2,500 scientists, economists, and policymakers from more than 120 countries, said that humans have been the primary cause of global warming since 1950. Humans have ways to cut greenhouse gas emissions and avoid the consequences of global warming, a major climate report concluded. But to change the climate, the transition from fossil fuels like coal and oil needs to occur within decades, according to the final report this year from the UN's Intergovernmental Panel on Climate Change (IPCC).

Some of the more common soil contaminants are chlorinated hydrocarbons (CFH), heavy metals (such as chromium, cadmium—found in rechargeable batteries, and lead—found in lead paint, aviation fuel and still in some countries, gasoline), MTBE, zinc, arsenic and benzene. In 2001 a series of press reports culminating in a book called *Fateful Harvest* unveiled a widespread practice of recycling industrial byproducts into fertilizer, resulting in the contamination of the soil with various metals. Ordinary municipal landfills are the source of many chemical substances entering the soil environment (and often groundwater), emanating from the wide variety of refuse accepted, especially substances illegally discarded there, or from pre-1970 landfills that may have been subject to little control in the U.S. or EU. There have also been some unusual releases of polychlorinated dibenzodioxins, commonly called *dioxins* for simplicity, such as TCDD.

Pollution can also be the consequence of a natural disaster. For example, hurricanes often involve water contamination from sewage, and petrochemical spills from ruptured boats or automobiles. Larger scale and environmental damage is not uncommon when coastal oil rigs or refineries are involved. Some sources of pollution, such as nuclear power plants or oil tankers, can produce widespread and potentially hazardous releases when accidents occur.

In the case of noise pollution the dominant source class is the motor vehicle, producing about ninety percent of all unwanted noise worldwide.

Effects

Human Health

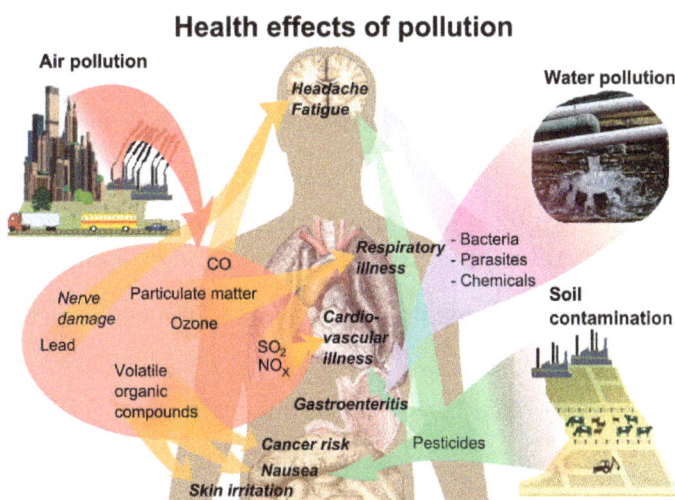

Overview of main health effects on humans from some common types of pollution.

Adverse air quality can kill many organisms including humans. Ozone pollution can cause respiratory disease, cardiovascular disease, throat inflammation, chest pain, and congestion. Water pollution causes approximately 14,000 deaths per day, mostly due to contamination of drinking water by untreated sewage in developing countries. An estimated 500 million Indians have no access to a proper toilet, Over ten million people in India fell ill with waterborne illnesses in 2013, and 1,535 people died, most of them children. Nearly 500 million Chinese lack access to safe drinking water. A 2010 analysis estimated that 1.2 million people died prematurely each year in China because of air pollution. The WHO estimated in 2007 that air pollution causes half a million deaths per year in India. Studies have estimated that the number of people killed annually in the United States could be over 50,000.

Oil spills can cause skin irritations and rashes. Noise pollution induces hearing loss, high blood pressure, stress, and sleep disturbance. Mercury has been linked to developmental deficits in children and neurologic symptoms. Older people are majorly exposed to diseases induced by air pollution. Those with heart or lung disorders are at additional risk. Children and infants are also at serious risk. Lead and other heavy metals have been shown to cause neurological problems. Chemical and radioactive substances can cause cancer and as well as birth defects.

Environment

Pollution has been found to be present widely in the environment. There are a number of effects of this:

- Biomagnification describes situations where toxins (such as heavy metals) may pass through trophic levels, becoming exponentially more concentrated in the process.

- Carbon dioxide emissions cause ocean acidification, the ongoing decrease in the pH of the Earth's oceans as CO_2 becomes dissolved.

- The emission of greenhouse gases leads to global warming which affects ecosystems in many ways.

- Invasive species can out compete native species and reduce biodiversity. Invasive plants can contribute debris and biomolecules (allelopathy) that can alter soil and chemical compositions of an environment, often reducing native species competitiveness.

Nitrogen oxides are removed from the air by rain and fertilise land which can change the species composition of ecosystems.

Smog and haze can reduce the amount of sunlight received by plants to carry out photosynthesis and leads to the production of tropospheric ozone which damages plants.

- Soil can become infertile and unsuitable for plants. This will affect other organisms in the food web.

- Sulfur dioxide and nitrogen oxides can cause acid rain which lowers the pH value of soil.

Environmental Health Information

The Toxicology and Environmental Health Information Program (TEHIP) at the United States National Library of Medicine (NLM) maintains a comprehensive toxicology and environmental health web site that includes access to resources produced by TEHIP and by other government agencies and organizations. This web site includes links to databases, bibliographies, tutorials, and other scientific and consumer-oriented resources. TEHIP also is responsible for the Toxicology

Data Network (TOXNET) an integrated system of toxicology and environmental health databases that are available free of charge on the web.

TOXMAP is a Geographic Information System (GIS) that is part of TOXNET. TOXMAP uses maps of the United States to help users visually explore data from the United States Environmental Protection Agency's (EPA) Toxics Release Inventory and Superfund Basic Research Programs.

Worker Productivity

A number of studies show that pollution has an adverse effect on the productivity of both indoor and outdoor workers.

Regulation and Monitoring

To protect the environment from the adverse effects of pollution, many nations worldwide have enacted legislation to regulate various types of pollution as well as to mitigate the adverse effects of pollution.

Pollution Control

A litter trap catches floating waste in the Yarra River, east-central Victoria, Australia

A dust collector in Pristina, Kosovo

Gas nozzle with vapor recovery

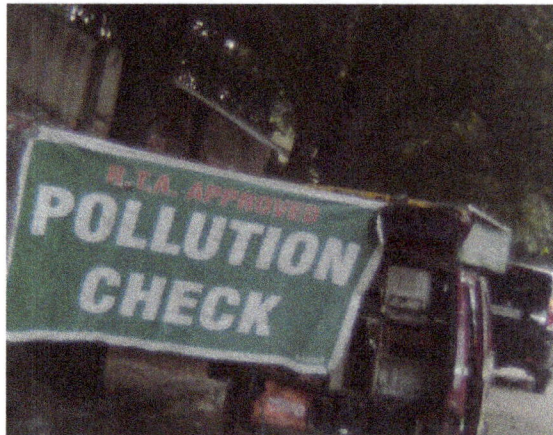

A Mobile Pollution Check Vehicle in India.

Pollution control is a term used in environmental management. It means the control of emissions and effluents into air, water or soil. Without pollution control, the waste products from overconsumption, heating, agriculture, mining, manufacturing, transportation and other human activities, whether they accumulate or disperse, will degrade the environment. In the hierarchy of controls, pollution prevention and waste minimization are more desirable than pollution control. In the field of land development, low impact development is a similar technique for the prevention of urban runoff.

Practices

- Recycling

- Reusing

- Waste minimisation

- Mitigating

- Preventing

- Compost

Pollution Control Devices

- Dust collection systems
  - o Baghouses
  - o Cyclones
  - o Electrostatic precipitators
- Scrubbers
  - o Baffle spray scrubber
  - o Cyclonic spray scrubber
  - o Ejector venturi scrubber
  - o Mechanically aided scrubber
  - o Spray tower
  - o Wet scrubber
- Sewage treatment
  - o Sedimentation (Primary treatment)
  - o Activated sludge biotreaters (Secondary treatment; also used for industrial wastewater)
  - o Aerated lagoons
  - o Constructed wetlands (also used for urban runoff)
- Industrial wastewater treatment
  - o API oil-water separators
  - o Biofilters
  - o Dissolved air flotation (DAF)
  - o Powdered activated carbon treatment
  - o Ultrafiltration
- Vapor recovery systems
- Phytoremediation

Perspectives

The earliest precursor of pollution generated by life forms would have been a natural function of their existence. The attendant consequences on viability and population levels fell within the sphere of natural selection. These would have included the demise of a population locally or ultimately, species extinction. Processes that were untenable would have resulted in a new balance brought about by changes and adaptations. At the extremes, for any form of life, consideration of pollution is superseded by that of survival.

For humankind, the factor of technology is a distinguishing and critical consideration, both as an enabler and an additional source of byproducts. Short of survival, human concerns include the range from quality of life to health hazards. Since science holds experimental demonstration to be definitive, modern treatment of toxicity or environmental harm involves defining a level at which an effect is observable. Common examples of fields where practical measurement is crucial include automobile emissions control, industrial exposure (e.g. Occupational Safety and Health Administration (OSHA) PELs), toxicology (e.g. LD_{50}), and medicine (e.g. medication and radiation doses).

"The solution to pollution is dilution", is a dictum which summarizes a traditional approach to pollution management whereby sufficiently diluted pollution is not harmful. It is well-suited to some other modern, locally scoped applications such as laboratory safety procedure and hazardous material release emergency management. But it assumes that the dilutant is in virtually unlimited supply for the application or that resulting dilutions are acceptable in all cases.

Such simple treatment for environmental pollution on a wider scale might have had greater merit in earlier centuries when physical survival was often the highest imperative, human population and densities were lower, technologies were simpler and their byproducts more benign. But these are often no longer the case. Furthermore, advances have enabled measurement of concentrations not possible before. The use of statistical methods in evaluating outcomes has given currency to the principle of probable harm in cases where assessment is warranted but resorting to deterministic models is impractical or infeasible. In addition, consideration of the environment beyond direct impact on human beings has gained prominence.

Yet in the absence of a superseding principle, this older approach predominates practices throughout the world. It is the basis by which to gauge concentrations of effluent for legal release, exceeding which penalties are assessed or restrictions applied. One such superseding principle is contained in modern hazardous waste laws in developed countries, as the process of diluting hazardous waste to make it non-hazardous is usually a regulated treatment process. Migration from pollution dilution to elimination in many cases can be confronted by challenging economical and technological barriers.

Greenhouse Gases and Global Warming

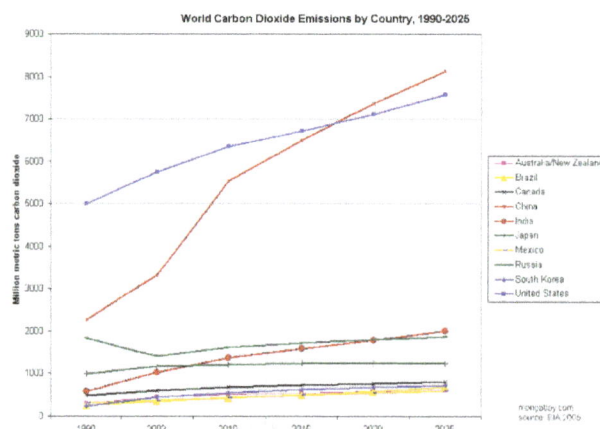

Historical and projected CO_2 emissions by country (as of 2005).
Source: Energy Information Administration.

Carbon dioxide, while vital for photosynthesis, is sometimes referred to as pollution, because raised levels of the gas in the atmosphere are affecting the Earth's climate. Disruption of the environment can also highlight the connection between areas of pollution that would normally be classified separately, such as those of water and air. Recent studies have investigated the potential for long-term rising levels of atmospheric carbon dioxide to cause slight but critical increases in the acidity of ocean waters, and the possible effects of this on marine ecosystems.

Most Polluted Places in the Developing World

The Blacksmith Institute, an international non-for-profit organization dedicated to eliminating life-threatening pollution in the developing world, issues an annual list of some of the world's worst polluted places.

References

- Beychok, Milton R. (1967). Aqueous Wastes from Petroleum and Petrochemical Plants (1st ed.). John Wiley & Sons. ISBN 0-471-07189-7. LCCN 67019834.

- Chang, Tom; Zivin, Joshua Graff; Gross, Tal; Neidell, Matthew (2016-06-01). "The Effect of Pollution on Worker Productivity: Evidence from Call-Center Workers in China". National Bureau of Economic Research.

- Li, Teng; Liu, Haoming; Salvo, Alberto (2015-05-29). "Severe Air Pollution and Labor Productivity". Rochester, NY: Social Science Research Network.

- Isalkar, Umesh (29 July 2014). "Over 1,500 lives lost to diarrhoea in 2013, delay in treatment blamed". The Times of India. Indiatimes. Retrieved 29 July 2014.

- The World's Most Polluted Places: The Top Ten of the Dirty Thirty (PDF), archived from the original (PDF) on 2007-10-11, retrieved 2013-12-10

- Jonathan Medalia, Comprehensive Nuclear-Test-Ban Treaty: Background and Current Developments (DIANE Publishing, 2013.)

- "The Mixture Rule under the Resource Conservation and Recovery Act" (PDF). U.S. Dept. of Energy. 1999. Retrieved 2012-04-10.

- Judith A. Layzer, "Love Canal: hazardous waste and politics of fear" in Layzer, The Environmental Case (CQ Press, 2012) pp: 56-82.

- Zivin, Joshua Graff; Neidell, Matthew (2012-12-01). "The Impact of Pollution on Worker Productivity". American Economic Review. 102 (7): 3652–3673. doi:10.1257/aer.102.7.3652. ISSN 0002-8282.

Major Types of Pollution

Pollution can be broadly classified into a few categories. This chapter looks at air pollution, water pollution, soil contamination and noise pollution. Classification also leads to better management and the development of environment-friendly technology that can put a cap on pollution.

Air Pollution

Air pollution from a fossil-fuel power station

Air pollution is the introduction of particulates, biological molecules, or other harmful materials into Earth's atmosphere, causing diseases, allergies, death to humans, damage to other living organisms such as animals and food crops, or the natural or built environment. Air pollution may come from anthropogenic or natural sources.

The atmosphere is a complex natural gaseous system that is essential to support life on planet Earth.

Indoor air pollution and urban air quality are listed as two of the world's worst toxic pollution problems in the 2008 Blacksmith Institute World's Worst Polluted Places report. According to the

2014 WHO report, air pollution in 2012 caused the deaths of around 7 million people worldwide, an estimate roughly matched by the International Energy Agency.

Pollutants

Carbon dioxide in Earth's atmosphere if *half* of global-warming emissions are *not* absorbed.
(NASA simulation; 9 November 2015)

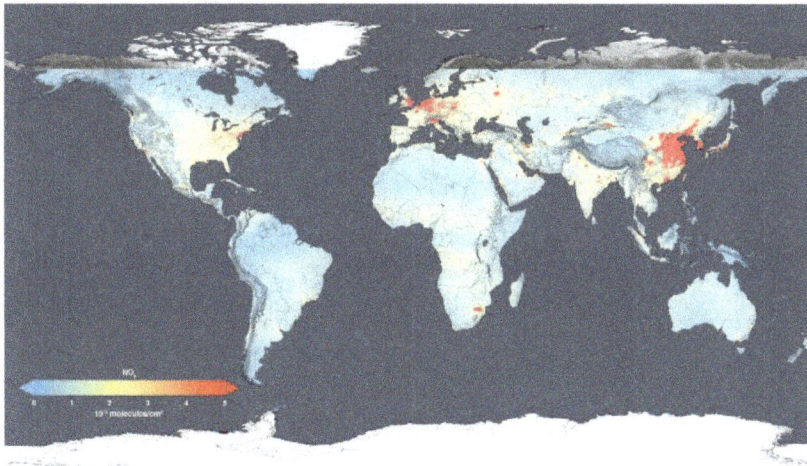

Nitrogen dioxide 2014 - global air quality levels
(released 14 December 2015).

An air pollutant is a substance in the air that can have adverse effects on humans and the ecosystem. The substance can be solid particles, liquid droplets, or gases. A pollutant can be of natural origin or man-made. Pollutants are classified as primary or secondary. Primary pollutants are usually produced from a process, such as ash from a volcanic eruption. Other examples include carbon monoxide gas from motor vehicle exhaust, or the sulfur dioxide released from factories. Secondary pollutants are not emitted directly. Rather, they form in the air when primary pollutants react or interact. Ground level ozone is a prominent example of a secondary pollutant. Some pollutants may be both primary and secondary: they are both emitted directly and formed from other primary pollutants.

Before flue-gas desulfurization was installed, the emissions from this power plant in New Mexico contained excessive amounts of sulfur dioxide.

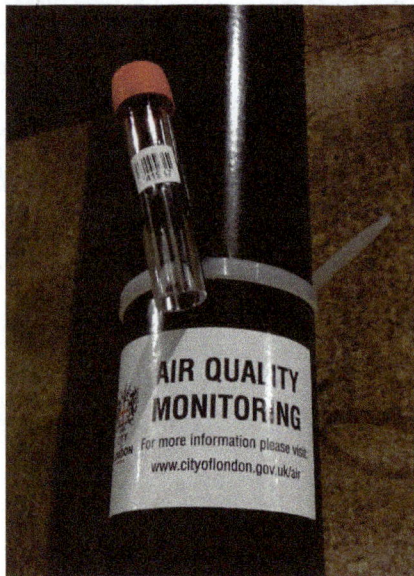

Nitrogen dioxide diffusion tube for air quality monitoring. Positioned in London City.

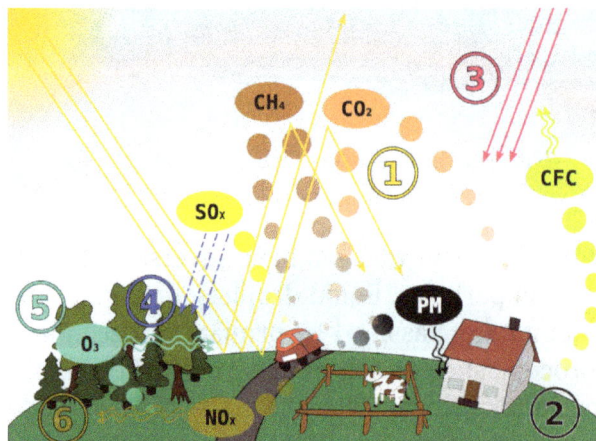

Schematic drawing, causes and effects of air pollution: (1) greenhouse effect, (2) particulate contamination, (3) increased UV radiation, (4) acid rain, (5) increased ground level ozone concentration, (6) increased levels of nitrogen oxides.

Major primary pollutants produced by human activity include:

- Sulfur oxides (SO_x) - particularly sulfur dioxide, a chemical compound with the formula SO_2. SO_2 is produced by volcanoes and in various industrial processes. Coal and petroleum often contain sulfur compounds, and their combustion generates sulfur dioxide. Further oxidation of SO_2, usually in the presence of a catalyst such as NO_2, forms H_2SO_4, and thus acid rain. This is one of the causes for concern over the environmental impact of the use of these fuels as power sources.

- Nitrogen oxides (NO_x) - Nitrogen oxides, particularly nitrogen dioxide, are expelled from high temperature combustion, and are also produced during thunderstorms by electric discharge. They can be seen as a brown haze dome above or a plume downwind of cities. Nitrogen dioxide is a chemical compound with the formula NO_2. It is one of several nitrogen oxides. One of the most prominent air pollutants, this reddish-brown toxic gas has a characteristic sharp, biting odor.

- Carbon monoxide (CO) - CO is a colorless, odorless, toxic yet non-irritating gas. It is a product by incomplete combustion of fuel such as natural gas, coal or wood. Vehicular exhaust is a major source of carbon monoxide.

- Volatile organic compounds (VOC) - VOCs are a well-known outdoor air pollutant. They are categorized as either methane (CH_4) or non-methane (NMVOCs). Methane is an extremely efficient greenhouse gas which contributes to enhanced global warming. Other hydrocarbon VOCs are also significant greenhouse gases because of their role in creating ozone and prolonging the life of methane in the atmosphere. This effect varies depending on local air quality. The aromatic NMVOCs benzene, toluene and xylene are suspected carcinogens and may lead to leukemia with prolonged exposure. 1,3-butadiene is another dangerous compound often associated with industrial use.

- Particulates, alternatively referred to as particulate matter (PM), atmospheric particulate matter, or fine particles, are tiny particles of solid or liquid suspended in a gas. In contrast, aerosol refers to combined particles and gas. Some particulates occur naturally, originating from volcanoes, dust storms, forest and grassland fires, living vegetation, and sea spray. Human activities, such as the burning of fossil fuels in vehicles, power plants and various industrial processes also generate significant amounts of aerosols. Averaged worldwide, anthropogenic aerosols—those made by human activities—currently account for approximately 10 percent of our atmosphere. Increased levels of fine particles in the air are linked to health hazards such as heart disease, altered lung function and lung cancer.

- Persistent free radicals connected to airborne fine particles are linked to cardiopulmonary disease.

- Toxic metals, such as lead and mercury, especially their compounds.

- Chlorofluorocarbons (CFCs) - harmful to the ozone layer; emitted from products are currently banned from use. These are gases which are released from air conditioners, refrigerators, aerosol sprays, etc. CFC's on being released into the air rises to stratosphere. Here they come in contact with other gases and damage the ozone layer. This allows harmful ultraviolet rays to reach the earth's surface. This can lead to skin cancer, disease to eye and can even cause damage to plants.

- Ammonia (NH_3) - emitted from agricultural processes. Ammonia is a compound with the formula NH_3. It is normally encountered as a gas with a characteristic pungent odor. Ammonia contributes significantly to the nutritional needs of terrestrial organisms by serving as a precursor to foodstuffs and fertilizers. Ammonia, either directly or indirectly, is also a building block for the synthesis of many pharmaceuticals. Although in wide use, ammonia is both caustic and hazardous. In the atmosphere, ammonia reacts with oxides of nitrogen and sulfur to form secondary particles.

- Odours — such as from garbage, sewage, and industrial processes

- Radioactive pollutants - produced by nuclear explosions, nuclear events, war explosives, and natural processes such as the radioactive decay of radon.

Secondary pollutants include:

- Particulates created from gaseous primary pollutants and compounds in photochemical smog. Smog is a kind of air pollution. Classic smog results from large amounts of coal burning in an area caused by a mixture of smoke and sulfur dioxide. Modern smog does not usually come from coal but from vehicular and industrial emissions that are acted on in the atmosphere by ultraviolet light from the sun to form secondary pollutants that also combine with the primary emissions to form photochemical smog.

- Ground level ozone (O_3) formed from NO_x and VOCs. Ozone (O_3) is a key constituent of the troposphere. It is also an important constituent of certain regions of the stratosphere commonly known as the Ozone layer. Photochemical and chemical reactions involving it drive many of the chemical processes that occur in the atmosphere by day and by night. At abnormally high concentrations brought about by human activities (largely the combustion of fossil fuel), it is a pollutant, and a constituent of smog.

- Peroxyacetyl nitrate (PAN) - similarly formed from NO_x and VOCs.

Minor air pollutants include:

- A large number of minor hazardous air pollutants. Some of these are regulated in USA under the Clean Air Act and in Europe under the Air Framework Directive

- A variety of persistent organic pollutants, which can attach to particulates

Persistent organic pollutants (POPs) are organic compounds that are resistant to environmental degradation through chemical, biological, and photolytic processes. Because of this, they have been observed to persist in the environment, to be capable of long-range transport, bioaccumulate in human and animal tissue, biomagnify in food chains, and to have potentially significant impacts on human health and the environment.

Sources

Dust storm approaching Stratford, Texas.

Controlled burning of a field outside of Statesboro, Georgia in preparation for spring planting.

There are various locations, activities or factors which are responsible for releasing pollutants into the atmosphere. These sources can be classified into two major categories.

Anthropogenic (Man-Made) Sources:

These are mostly related to the burning of multiple types of fuel.

- Stationary sources include smoke stacks of power plants, manufacturing facilities (factories) and waste incinerators, as well as furnaces and other types of fuel-burning heating devices. In developing and poor countries, traditional biomass burning is the major source of air pollutants; traditional biomass includes wood, crop waste and dung.

- Mobile sources include motor vehicles, marine vessels, and aircraft.

- Controlled burn practices in agriculture and forest management. Controlled or prescribed burning is a technique sometimes used in forest management, farming, prairie restoration or greenhouse gas abatement. Fire is a natural part of both forest and grassland ecology and controlled fire can be a tool for foresters. Controlled burning stimulates the germination of some desirable forest trees, thus renewing the forest.

- Fumes from paint, hair spray, varnish, aerosol sprays and other solvents

- Waste deposition in landfills, which generate methane. Methane is highly flammable and may form explosive mixtures with air. Methane is also an asphyxiant and may displace oxygen in an enclosed space. Asphyxia or suffocation may result if the oxygen concentration is reduced to below 19.5% by displacement.

- Military resources, such as nuclear weapons, toxic gases, germ warfare and rocketry

Natural Sources:

- Dust from natural sources, usually large areas of land with little or no vegetation

- Methane, emitted by the digestion of food by animals, for example cattle

- Radon gas from radioactive decay within the Earth's crust. Radon is a colorless, odorless, naturally occurring, radioactive noble gas that is formed from the decay of radium. It is considered to be a health hazard. Radon gas from natural sources can accumulate in buildings, especially in confined areas such as the basement and it is the second most frequent cause of lung cancer, after cigarette smoking.

- Smoke and carbon monoxide from wildfires

- Vegetation, in some regions, emits environmentally significant amounts of Volatile organic compounds (VOCs) on warmer days. These VOCs react with primary anthropogenic pollutants—specifically, NO_x, SO_2, and anthropogenic organic carbon compounds — to produce a seasonal haze of secondary pollutants. Black gum, poplar, oak and willow are some examples of vegetation that can produce abundant VOCs. The VOC production from these species result in ozone levels up to eight times higher than the low-impact tree species.

- Volcanic activity, which produces sulfur, chlorine, and ash particulates

Emission Factors

Beijing air on a 2005-day after rain (left) and a smoggy day (right)

Air pollutant emission factors are reported representative values that attempt to relate the quantity of a pollutant released to the ambient air with an activity associated with the release of that pollutant. These factors are usually expressed as the weight of pollutant divided by a unit weight, volume, distance, or duration of the activity emitting the pollutant (e.g., kilograms of particulate emitted per tonne of coal burned). Such factors facilitate estimation of emissions from various sources of air pollution. In most cases, these factors are simply averages of all available data of acceptable quality, and are generally assumed to be representative of long-term averages.

There are 12 compounds in the list of Persistent organic pollutants. Dioxins and furans are two of them and intentionally created by combustion of organics, like open burning of plastics. These compounds are also endocrine disruptors and can mutate the human genes.

The United States Environmental Protection Agency has published a compilation of air pollutant emission factors for a multitude of industrial sources. The United Kingdom, Australia, Canada and many other countries have published similar compilations, as well as the European Environment Agency.

Air Pollution Exposure

Air pollution risk is a function of the hazard of the pollutant and the exposure to that pollutant. Air pollution exposure can be expressed for an individual, for certain groups (e.g. neighborhoods or children living in a country), or for entire populations. For example, one may want to calculate the exposure to a hazardous air pollutant for a geographic area, which includes the various micro-environments and age groups. This can be calculated as an inhalation exposure. This would account for daily exposure in various settings (e.g. different indoor micro-environments and outdoor locations). The exposure needs to include different age and other demographic groups, especially infants, children, pregnant women and other sensitive subpopulations. The exposure to an air pollutant must integrate the concentrations of the air pollutant with respect to the time spent in each setting and the respective inhalation rates for each subgroup for each specific time that the subgroup is in the setting and engaged in particular activities (playing, cooking, reading, working, etc.). For example, a small child's inhalation rate will be less than that of an adult. A child engaged in vigorous exercise will have a higher respiration rate than the same child in a sedentary activity. The daily exposure, then, needs to reflect the time spent in each micro-environmental setting and the type of activities in these settings. The air pollutant concentration in each microactivity/ microenvironmental setting is summed to indicate the exposure.

Indoor Air Quality (IAQ)

Air quality monitoring, New Delhi, India.

A lack of ventilation indoors concentrates air pollution where people often spend the majority of their time. Radon (Rn) gas, a carcinogen, is exuded from the Earth in certain locations and trapped inside houses. Building materials including carpeting and plywood emit formaldehyde (H_2CO) gas. Paint and solvents give off volatile organic compounds (VOCs) as they dry. Lead paint can degenerate into dust and be inhaled. Intentional air pollution is introduced with the use of air fresheners, incense, and other scented items. Controlled wood fires in stoves and fireplaces can add significant amounts of smoke particulates into the air, inside and out. Indoor pollution fatalities may be caused by using pesticides and other chemical sprays indoors without proper ventilation.

Carbon monoxide poisoning and fatalities are often caused by faulty vents and chimneys, or by the burning of charcoal indoors or in a confined space, such as a tent. Chronic carbon monoxide poisoning can result even from poorly-adjusted pilot lights. Traps are built into all domestic plumbing to keep sewer gas and hydrogen sulfide, out of interiors. Clothing emits tetrachloroethylene, or other dry cleaning fluids, for days after dry cleaning.

Though its use has now been banned in many countries, the extensive use of asbestos in industrial and domestic environments in the past has left a potentially very dangerous material in many localities. Asbestosis is a chronic inflammatory medical condition affecting the tissue of the lungs. It occurs after long-term, heavy exposure to asbestos from asbestos-containing materials in structures. Sufferers have severe dyspnea (shortness of breath) and are at an increased risk regarding several different types of lung cancer. As clear explanations are not always stressed in non-technical literature, care should be taken to distinguish between several forms of relevant diseases. According to the World Health Organisation (WHO), these may defined as; asbestosis, *lung cancer*, and *Peritoneal Mesothelioma* (generally a very rare form of cancer, when more widespread it is almost always associated with prolonged exposure to asbestos).

Biological sources of air pollution are also found indoors, as gases and airborne particulates. Pets produce dander, people produce dust from minute skin flakes and decomposed hair, dust mites in bedding, carpeting and furniture produce enzymes and micrometre-sized fecal droppings, inhabitants emit methane, mold forms on walls and generates mycotoxins and spores, air conditioning systems can incubate Legionnaires' disease and mold, and houseplants, soil and surrounding gardens can produce pollen, dust, and mold. Indoors, the lack of air circulation allows these airborne pollutants to accumulate more than they would otherwise occur in nature.

Health Effects

Air pollution is a significant risk factor for a number of pollution-related diseases and health conditions including respiratory infections, heart disease, COPD, stroke and lung cancer. The health effects caused by air pollution may include difficulty in breathing, wheezing, coughing, asthma and worsening of existing respiratory and cardiac conditions. These effects can result in increased medication use, increased doctor or emergency room visits, more hospital admissions and premature death. The human health effects of poor air quality are far reaching, but principally affect the body's respiratory system and the cardiovascular system. Individual reactions to air pollutants depend on the type of pollutant a person is exposed to, the degree of exposure, and the individual's health status and genetics. The most common sources of air pollution include particulates, ozone, nitrogen dioxide, and sulphur dioxide. Children aged less than five years that live in developing countries are the most vulnerable population in terms of total deaths attributable to indoor and outdoor air pollution.

Mortality

The World Health Organization estimated in 2014 that every year air pollution causes the premature death of some 7 million people worldwide. India has the highest death rate due to air pollution. India also has more deaths from asthma than any other nation according to the World Health Organization. In December 2013 air pollution was estimated to kill 500,000 people in China each year. There is a positive correlation between pneumonia-related deaths and air pollution from motor vehicle emissions.

Annual premature European deaths caused by air pollution are estimated at 430,000. An important cause of these deaths is nitrogen dioxide and other nitrogen oxides (NOx) emitted by road vehicles. Across the European Union, air pollution is estimated to reduce life expectancy by almost nine months. Causes of deaths include strokes, heart disease, COPD, lung cancer, and lung infections.

The US EPA estimates that a proposed set of changes in diesel engine technology (*Tier 2*) could result in 12,000 fewer *premature mortalities*, 15,000 fewer heart attacks, 6,000 fewer emergency room visits by children with asthma, and 8,900 fewer respiratory-related hospital admissions each year in the United States.

The US EPA has estimated that limiting ground-level ozone concentration to 65 parts per billion, would avert 1,700 to 5,100 premature deaths nationwide in 2020 compared with the 75-ppb standard. The agency projected the more protective standard would also prevent an additional 26,000 cases of aggravated asthma, and more than a million cases of missed work or school. Following this assessment, the EPA acted to protect public health by lowering the National Ambient Air Quality Standards (NAAQS) for ground-level ozone to 70 parts per billion (ppb).

A new economic study of the health impacts and associated costs of air pollution in the Los Angeles Basin and San Joaquin Valley of Southern California shows that more than 3,800 people die prematurely (approximately 14 years earlier than normal) each year because air pollution levels violate federal standards. The number of annual premature deaths is considerably higher than the fatalities related to auto collisions in the same area, which average fewer than 2,000 per year.

Diesel exhaust (DE) is a major contributor to combustion-derived particulate matter air pollution. In several human experimental studies, using a well-validated exposure chamber setup, DE has been linked to acute vascular dysfunction and increased thrombus formation.

The mechanisms linking air pollution to increased cardiovascular mortality are uncertain, but probably include pulmonary and systemic inflammation.

Cardiovascular Disease

A 2007 review of evidence found ambient air pollution exposure is a risk factor correlating with increased total mortality from cardiovascular events (range: 12% to 14% per 10 microg/m³ increase).

Air pollution is also emerging as a risk factor for stroke, particularly in developing countries where pollutant levels are highest. A 2007 study found that in women, air pollution is not associated with hemorrhagic but with ischemic stroke. Air pollution was also found to be associated with increased incidence and mortality from coronary stroke in a cohort study in 2011. Associations are believed to be causal and effects may be mediated by vasoconstriction, low-grade inflammation and atherosclerosis Other mechanisms such as autonomic nervous system imbalance have also been suggested.

Lung Disease

Chronic obstructive pulmonary disease (COPD) includes diseases such as chronic bronchitis and emphysema.

Research has demonstrated increased risk of developing asthma and COPD from increased exposure to traffic-related air pollution. Additionally, air pollution has been associated with increased hospitalization and mortality from asthma and COPD.

A study conducted in 1960-1961 in the wake of the Great Smog of 1952 compared 293 London residents with 477 residents of Gloucester, Peterborough, and Norwich, three towns with low reported death rates from chronic bronchitis. All subjects were male postal truck drivers aged 40 to 59. Compared to the subjects from the outlying towns, the London subjects exhibited more severe respiratory symptoms (including cough, phlegm, and dyspnea), reduced lung function (FEV_1 and peak flow rate), and increased sputum production and purulence. The differences were more pronounced for subjects aged 50 to 59. The study controlled for age and smoking habits, so concluded that air pollution was the most likely cause of the observed differences.

It is believed that much like cystic fibrosis, by living in a more urban environment serious health hazards become more apparent. Studies have shown that in urban areas patients suffer mucus hypersecretion, lower levels of lung function, and more self-diagnosis of chronic bronchitis and emphysema.

Cancer

Cancer mainly the result of environmental factors.

A review of evidence regarding whether ambient air pollution exposure is a risk factor for cancer in 2007 found solid data to conclude that long-term exposure to PM2.5 (fine particulates) increases the overall risk of non-accidental mortality by 6% per a 10 microg/m³ increase. Exposure to PM2.5 was also associated with an increased risk of mortality from lung cancer (range: 15% to 21% per 10 microg/m³ increase) and total cardiovascular mortality (range: 12% to 14% per a 10 microg/m³ increase). The review further noted that living close to busy traffic appears to be associated with elevated risks of these three outcomes --- increase in lung cancer deaths, cardiovascular deaths, and overall non-accidental deaths. The reviewers also found suggestive evidence that exposure to PM2.5 is positively associated with mortality from coronary heart diseases and exposure to SO_2 increases mortality from lung cancer, but the data was insufficient to provide solid conclusions. Another investigation showed that higher activity level increases deposition fraction of aerosol particles in human lung and recommended avoiding heavy activities like running in outdoor space at polluted areas.

In 2011, a large Danish epidemiological study found an increased risk of lung cancer for patients who lived in areas with high nitrogen oxide concentrations. In this study, the association was higher for non-smokers than smokers. An additional Danish study, also in 2011, likewise noted evidence of possible associations between air pollution and other forms of cancer, including cervical cancer and brain cancer.

In December 2015, medical scientists reported that cancer is overwhelmingly a result of environmental factors, and not largely down to bad luck. Maintaining a healthy weight, eating a healthy diet, minimizing alcohol and eliminating smoking reduces the risk of developing the disease, according to the researchers.

Children

In the United States, despite the passage of the Clean Air Act in 1970, in 2002 at least 146 million Americans were living in non-attainment areas—regions in which the concentration of certain air pollutants exceeded federal standards. These dangerous pollutants are known as the criteria pollutants, and include ozone, particulate matter, sulfur dioxide, nitrogen dioxide, carbon monoxide, and lead. Protective measures to ensure children's health are being taken in cities such as New Delhi, India where buses now use compressed natural gas to help eliminate the "pea-soup" smog. A recent study in Europe has found that exposure to ultrafine particles can increase blood pressure in children.

"Clean" Areas

Even in the areas with relatively low levels of air pollution, public health effects can be significant and costly, since a large number of people breathe in such pollutants. A 2005 scientific study for the British Columbia Lung Association showed that a small improvement in air quality (1% reduction of ambient PM2.5 and ozone concentrations) would produce $29 million in annual savings in the Metro Vancouver region in 2010. This finding is based on health valuation of lethal (death) and sub-lethal (illness) affects.

Central Nervous System

Data is accumulating that air pollution exposure also affects the central nervous system.

In a June 2014 study conducted by researchers at the University of Rochester Medical Center, published in the journal Environmental Health Perspectives, it was discovered that early exposure to air pollution causes the same damaging changes in the brain as autism and schizophrenia. The study also shows that air pollution also affected short-term memory, learning ability, and impulsivity. Lead researcher Professor Deborah Cory-Slechta said that "When we looked closely at the ventricles, we could see that the white matter that normally surrounds them hadn't fully developed. It appears that inflammation had damaged those brain cells and prevented that region of the brain from developing, and the ventricles simply expanded to fill the space. Our findings add to the growing body of evidence that air pollution may play a role in autism, as well as in other neurodevelopmental disorders." Air pollution has a more significant negative effect of males than on females.

In 2015, experimental studies reported the detection of significant episodic (situational) cognitive impairment from impurities in indoor air breathed by test subjects who were not informed about changes in the air quality. Researchers at the Harvard University and SUNY Upstate Medical University and Syracuse University measured the cognitive performance of 24 participants in three different controlled laboratory atmospheres that simulated those found in "conventional" and "green" buildings, as well as green buildings with enhanced ventilation. Performance was evaluated objectively using the widely used Strategic Management Simulation software simulation tool, which is a well-validated assessment test for executive decision-making in an unconstrained situation allowing initiative and improvisation. Significant deficits were observed in the performance scores achieved in increasing concentrations of either volatile organic compounds (VOCs) or carbon dioxide, while keeping other factors constant. The highest impurity levels reached are not uncommon in some classroom or office environments.

Agricultural Effects

In India in 2014, it was reported that air pollution by black carbon and ground level ozone had cut crop yields in the most affected areas by almost half in 2010 when compared to 1980 levels.

Historical Disasters

The world's worst short-term civilian pollution crisis was the 1984 Bhopal Disaster in India. Leaked industrial vapours from the Union Carbide factory, belonging to Union Carbide, Inc., U.S.A. (later bought by Dow Chemical Company), killed at least 3787 people and injured anywhere from 150,000 to 600,000. The United Kingdom suffered its worst air pollution event when the December 4 Great Smog of 1952 formed over London. In six days more than 4,000 died and more recent estimates put the figure at nearer 12,000. An accidental leak of anthrax spores from a biological warfare laboratory in the former USSR in 1979 near Sverdlovsk is believed to have caused at least 64 deaths. The worst single incident of air pollution to occur in the US occurred in Donora, Pennsylvania in late October, 1948, when 20 people died and over 7,000 were injured.

Alternatives to Avoid

There are now practical alternatives to the three principal causes of air pollution.

- Combustion of fossil fuels for space heating can be replaced by using ground source heat pumps and seasonal thermal energy storage.

- Electric power generation from burning fossil fuels can be replaced by power generation from nuclear and renewables.

- Motor vehicles driven by fossil fuels, a key factor in urban air pollution, can be replaced by electric vehicles.

Reduction Efforts

There are various air pollution control technologies and strategies available to reduce air pollution. At its most basic level, land-use planning is likely to involve zoning and transport infrastructure planning. In most developed countries, land-use planning is an important part of social policy, ensuring that land is used efficiently for the benefit of the wider economy and population, as well as to protect the environment.

Because a large share of air pollution is caused by combustion of fossil fuels such as coal and oil, the reduction of these fuels can reduce air pollution drastically. Most effective is the switch to clean power sources such as wind power, solar power, hydro power which don't cause air pollution. Efforts to reduce pollution from mobile sources includes primary regulation (many developing countries have permissive regulations), expanding regulation to new sources (such as cruise and transport ships, farm equipment, and small gas-powered equipment such as string trimmers, chainsaws, and snowmobiles), increased fuel efficiency (such as through the use of hybrid vehicles), conversion to cleaner fuels or conversion to electric vehicles.

Titanium dioxide has been researched for its ability to reduce air pollution. Ultraviolet light will release free electrons from material, thereby creating free radicals, which break up VOCs and NOx gases. One form is superhydrophilic.

In 2014, Prof. Tony Ryan and Prof. Simon Armitage of University of Sheffield prepared a 10 meter by 20 meter-sized poster coated with microscopic, pollution-eating nanoparticles of titanium dioxide. Placed on a building, this giant poster can absorb the toxic emission from around 20 cars each day.

A very effective means to reduce air pollution is the transition to renewable energy. According to a study published in Energy and Environmental Science in 2015 the switch to 100% renewable energy in the United States would eliminate about 62,000 premature mortalities per year and about 42,000 in 2050, if no biomass were used. This would save about $600 billion in health costs a year due to reduced air pollution in 2050, or about 3.6% of the 2014 U.S. gross domestic product.

Control Devices

The following items are commonly used as pollution control devices in industry and transportation. They can either destroy contaminants or remove them from an exhaust stream before it is emitted into the atmosphere.

- Particulate Control
  - o Mechanical collectors (dust cyclones, multicyclones)
  - o Electrostatic precipitators An electrostatic precipitator (ESP), or electrostatic air cleaner is a particulate collection device that removes particles from a flowing gas (such as

air), using the force of an induced electrostatic charge. Electrostatic precipitators are highly efficient filtration devices that minimally impede the flow of gases through the device, and can easily remove fine particulates such as dust and smoke from the air stream.

o Baghouses Designed to handle heavy dust loads, a dust collector consists of a blower, dust filter, a filter-cleaning system, and a dust receptacle or dust removal system (distinguished from air cleaners which utilize disposable filters to remove the dust).

o Particulate scrubbers Wet scrubber is a form of pollution control technology. The term describes a variety of devices that use pollutants from a furnace flue gas or from other gas streams. In a wet scrubber, the polluted gas stream is brought into contact with the scrubbing liquid, by spraying it with the liquid, by forcing it through a pool of liquid, or by some other contact method, so as to remove the pollutants.

- Scrubbers
  o Baffle spray scrubber
  o Cyclonic spray scrubber
  o Ejector venturi scrubber
  o Mechanically aided scrubber
  o Spray tower
  o Wet scrubber

- NOx control
  o Low NOx burners
  o Selective catalytic reduction (SCR)
  o Selective non-catalytic reduction (SNCR)
  o NOx scrubbers
  o Exhaust gas recirculation
  o Catalytic converter (also for VOC control)

- VOC abatement
  o Adsorption systems, using activated carbon, such as Fluidized Bed Concentrator
  o Flares
  o Thermal oxidizers
  o Catalytic converters
  o Biofilters
  o Absorption (scrubbing)
  o Cryogenic condensers
  o Vapor recovery systems

- Acid Gas/SO_2 control
  o Wet scrubbers

- o   Dry scrubbers
- o   Flue-gas desulfurization
- Mercury control
  - o   Sorbent Injection Technology
  - o   Electro-Catalytic Oxidation (ECO)
  - o   K-Fuel
- Dioxin and furan control
- Miscellaneous associated equipment
  - o   Source capturing systems
  - o   Continuous emissions monitoring systems (CEMS)

Regulations

Smog in Cairo

In general, there are two types of air quality standards. The first class of standards (such as the U.S. National Ambient Air Quality Standards and E.U. Air Quality Directive) set maximum atmospheric concentrations for specific pollutants. Environmental agencies enact regulations which are intended to result in attainment of these target levels. The second class (such as the North American Air Quality Index) take the form of a scale with various thresholds, which is used to communicate to the public the relative risk of outdoor activity. The scale may or may not distinguish between different pollutants.

Canada

In Canada, air pollution and associated health risks are measured with the Air Quality Health Index or (AQHI). It is a health protection tool used to make decisions to reduce short-term exposure to air pollution by adjusting activity levels during increased levels of air pollution.

The Air Quality Health Index or "AQHI" is a federal program jointly coordinated by Health Canada and Environment Canada. However, the AQHI program would not be possible without the commitment and support of the provinces, municipalities and NGOs. From air quality monitoring to health risk communication and community engagement, local partners are responsible for the vast majority of work related to AQHI implementation. The AQHI provides a number from 1 to 10+ to indicate the level of health risk associated with local air quality. Occasionally, when the amount of air pollution is abnormally high, the number may exceed 10. The AQHI provides a local air quality current value as well as a local air quality maximums forecast for today, tonight and tomorrow and provides associated health advice.

| 1 | 2 | 3 | 4 | 5 | 6 | 7 | 8 | 9 | 10 | + |

Risk: Low (1-3) Moderate (4-6) High (7-10) Very high (above 10)

As it is now known that even low levels of air pollution can trigger discomfort for the sensitive population, the index has been developed as a continuum: The higher the number, the greater the health risk and need to take precautions. The index describes the level of health risk associated with this number as 'low', 'moderate', 'high' or 'very high', and suggests steps that can be taken to reduce exposure.

Health Risk	Air Quality Health Index	Health Messages	
		At Risk population	General Population
Low	1-3	Enjoy your usual outdoor activities.	Ideal air quality for outdoor activities
Moderate	4-6	Consider reducing or rescheduling strenuous activities outdoors if you are experiencing symptoms.	No need to modify your usual outdoor activities unless you experience symptoms such as coughing and throat irritation.
High	7-10	Reduce or reschedule strenuous activities outdoors. Children and the elderly should also take it easy.	Consider reducing or rescheduling strenuous activities outdoors if you experience symptoms such as coughing and throat irritation.
Very high	Above 10	Avoid strenuous activities outdoors. Children and the elderly should also avoid outdoor physical exertion.	Reduce or reschedule strenuous activities outdoors, especially if you experience symptoms such as coughing and throat irritation.

The measurement is based on the observed relationship of Nitrogen Dioxide (NO_2), ground-level Ozone (O_3) and particulates ($PM_{2.5}$) with mortality, from an analysis of several Canadian cities. Significantly, all three of these pollutants can pose health risks, even at low levels of exposure, especially among those with pre-existing health problems.

When developing the AQHI, Health Canada's original analysis of health effects included five major air pollutants: particulates, ozone, and nitrogen dioxide (NO2), as well as sulfur dioxide (SO_2), and carbon monoxide (CO). The latter two pollutants provided little information in predicting health effects and were removed from the AQHI formulation.

The AQHI does not measure the effects of odour, pollen, dust, heat or humidity.

Germany

TA Luft is the German air quality regulation.

Hotspots

Air pollution hotspots are areas where air pollution emissions expose individuals to increased negative health effects. They are particularly common in highly populated, urban areas, where there may be a combination of stationary sources (e.g. industrial facilities) and mobile sources (e.g. cars and trucks) of pollution. Emissions from these sources can cause respiratory disease, childhood asthma, cancer, and other health problems. Fine particulate matter such as diesel soot, which contributes to more than 3.2 million premature deaths around the world each year, is a significant problem. It is very small and can lodge itself within the lungs and enter the bloodstream. Diesel soot is concentrated in densely populated areas, and one in six people in the U.S. live near a diesel pollution hot spot.

While air pollution hotspots affect a variety of populations, some groups are more likely to be located in hotspots. Previous studies have shown disparities in exposure to pollution by race and/or income. Hazardous land uses (toxic storage and disposal facilities, manufacturing facilities, major roadways) tend to be located where property values and income levels are low. Low socioeconomic status can be a proxy for other kinds of social vulnerability, including race, a lack of ability to influence regulation and a lack of ability to move to neighborhoods with less environmental pollution. These communities bear a disproportionate burden of environmental pollution and are more likely to face health risks such as cancer or asthma.

Studies show that patterns in race and income disparities not only indicate a higher exposure to pollution but also higher risk of adverse health outcomes. Communities characterized by low socioeconomic status and racial minorities can be more vulnerable to cumulative adverse health impacts resulting from elevated exposure to pollutants than more privileged communities. Blacks and Latinos generally face more pollution than whites and Asians, and low-income communities bear a higher burden of risk than affluent ones. Racial discrepancies are particularly distinct in suburban areas of the US South and metropolitan areas of the US West. Residents in public housing, who are generally low-income and cannot move to healthier neighborhoods, are highly affected by nearby refineries and chemical plants.

Cities

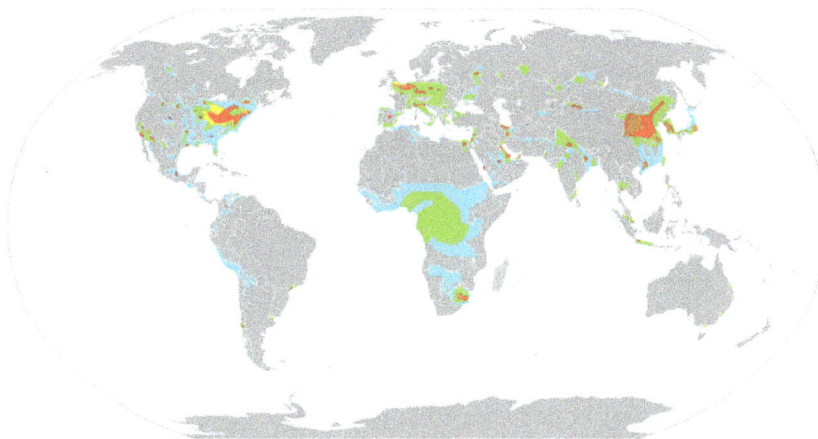

Nitrogen dioxide concentrations as measured from satellite 2002-2004

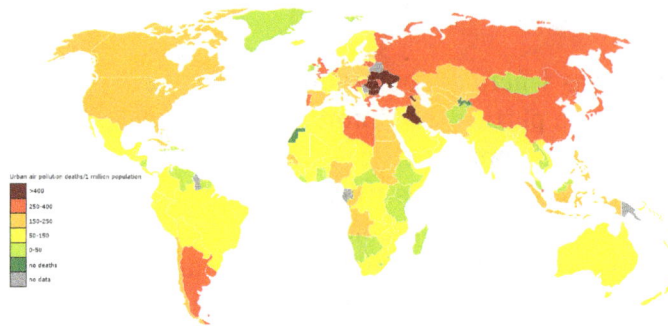

Deaths from air pollution in 2004

Air pollution is usually concentrated in densely populated metropolitan areas, especially in developing countries where environmental regulations are relatively lax or nonexistent. However, even populated areas in developed countries attain unhealthy levels of pollution, with Los Angeles and Rome being two examples. Between 2002 and 2011 the incidence of lung cancer in Beijing near doubled. While smoking remains the leading cause of lung cancer in China, the number of smokers is falling while lung cancer rates are rising.

National-scale Air Toxics Assessments 1995-2005

The national-scale air toxics assessment(NATA) is an evaluation of air toxics by the U.S. EPA. EPA has furnished four assessments that characterize nationwide chronic cancer risk estimates and noncancer hazards from inhaling air toxics. The lates was from 2005, and made publicly available in early 2011.

"EPA developed the NATA as a state-of-the-science screening tool for State/Local/Tribal Agencies to prioritize pollutants, emission sources and locations of interest for further study, in order to gain a better understanding of the risks. NATA assessments do not incorporate refined information about emission sources, but rather, use general information about sources to develop estimates of risks which are more likely to overestimate impacts than underestimate them. NATA provides estimates of the risk of cancer and other serious health effects from breathing (inhaling) air toxics in order to inform both national and more localized efforts to identify and prioritize air toxics, emission source types and locations which are of greatest potential concern in terms of contributing to population risk. This in turn helps air pollution experts focus limited analytical resources on areas and or populations where the potential for health risks are highest. Assessments include estimates of cancer and non-cancer health effects based on chronic exposure from outdoor sources, including assessments of non-cancer health effects for Diesel Particulate Matter. Assessments provide a snapshot of the outdoor air quality and the risks to human health that would result if air toxic emissions levels remained unchanged."

Most polluted cities by PM	
Particulate matter, μg/m³ (2004)	City
168	Cairo, Egypt
150	Delhi, India

128	Kolkata, India (Calcutta)
125	Tianjin, China
123	Chongqing, China
109	Kanpur, India
109	Lucknow, India
104	Jakarta, Indonesia
101	Shenyang, China

Governing Urban Air Pollution

In Europe, Council Directive 96/62/EC on ambient air quality assessment and management provides a common strategy against which member states can "set objectives for ambient air quality in order to avoid, prevent or reduce harmful effects on human health and the environment . . . and improve air quality where it is unsatisfactory".

On 25 July 2008 in the case Dieter Janecek v Freistaat Bayern CURIA, the European Court of Justice ruled that under this directive citizens have the right to require national authorities to implement a short term action plan that aims to maintain or achieve compliance to air quality limit values.

This important case law appears to confirm the role of the EC as centralised regulator to European nation-states as regards air pollution control. It places a supranational legal obligation on the UK to protect its citizens from dangerous levels of air pollution, furthermore superseding national interests with those of the citizen.

In 2010, the European Commission (EC) threatened the UK with legal action against the successive breaching of PM10 limit values. The UK government has identified that if fines are imposed, they could cost the nation upwards of £300 million per year.

In March 2011, the Greater London Built-up Area remains the only UK region in breach of the EC's limit values, and has been given 3 months to implement an emergency action plan aimed at meeting the EU Air Quality Directive. The City of London has dangerous levels of PM10 concentrations, estimated to cause 3000 deaths per year within the city. As well as the threat of EU fines, in 2010 it was threatened with legal action for scrapping the western congestion charge zone, which is claimed to have led to an increase in air pollution levels.

In response to these charges, Boris Johnson, Mayor of London, has criticised the current need for European cities to communicate with Europe through their nation state's central government, arguing that in future "A great city like London" should be permitted to bypass its government and deal directly with the European Commission regarding its air quality action plan.

This can be interpreted as recognition that cities can transcend the traditional national government organisational hierarchy and develop solutions to air pollution using global governance networks, for example through transnational relations. Transnational relations include but are not exclusive to national governments and intergovernmental organisations, allowing sub-national actors including cities and regions to partake in air pollution control as independent actors.

Particularly promising at present are global city partnerships. These can be built into networks, for example the C40 Cities Climate Leadership Group, of which London is a member. The C40 is a public 'non-state' network of the world's leading cities that aims to curb their greenhouse emissions. The C40 has been identified as 'governance from the middle' and is an alternative to intergovernmental policy. It has the potential to improve urban air quality as participating cities "exchange information, learn from best practices and consequently mitigate carbon dioxide emissions independently from national government decisions". A criticism of the C40 network is that its exclusive nature limits influence to participating cities and risks drawing resources away from less powerful city and regional actors.

Atmospheric Dispersion

The basic technology for analyzing air pollution is through the use of a variety of mathematical models for predicting the transport of air pollutants in the lower atmosphere. The principal methodologies are:

- Point source dispersion, used for industrial sources

- Line source dispersion, used for airport and roadway air dispersion modeling

- Area source dispersion, used for forest fires or duststorms

- Photochemical models, used to analyze reactive pollutants that form smog

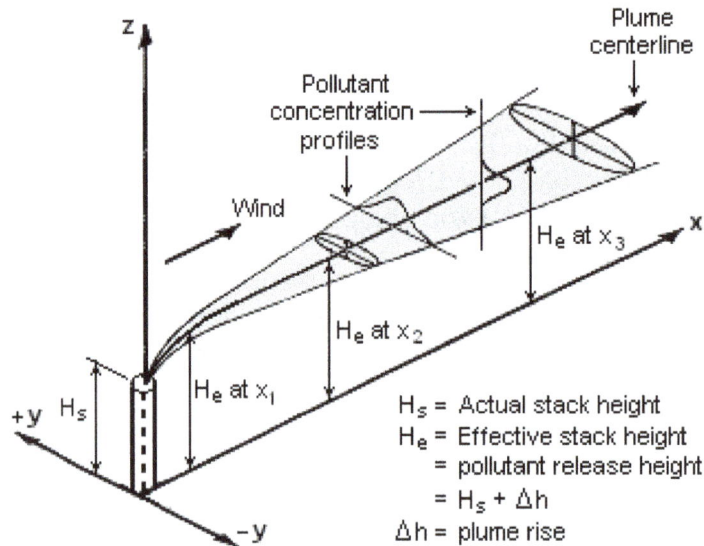

Visualization of a buoyant Gaussian air pollution dispersion plume as used in many atmospheric dispersion models.

The point source problem is the best understood, since it involves simpler mathematics and has been studied for a long period of time, dating back to about the year 1900. It uses a Gaussian dispersion model for continuous buoyant pollution plumes to predict the air pollution isopleths, with consideration given to wind velocity, stack height, emission rate and stability class (a measure of atmospheric turbulence). This model has been extensively validated and calibrated with experimental data for all sorts of atmospheric conditions.

The roadway air dispersion model was developed starting in the late 1950s and early 1960s in response to requirements of the National Environmental Policy Act and the U.S. Department of Transportation (then known as the Federal Highway Administration) to understand impacts of proposed new highways upon air quality, especially in urban areas. Several research groups were active in this model development, among which were: the Environmental Research and Technology (ERT) group in Lexington, Massachusetts, the ESL Inc. group in Sunnyvale, California and the California Air Resources Board group in Sacramento, California. The research of the ESL group received a boost with a contract award from the United States Environmental Protection Agency to validate a line source model using sulfur hexafluoride as a tracer gas. This program was successful in validating the line source model developed by ESL Inc. Some of the earliest uses of the model were in court cases involving highway air pollution; the Arlington, Virginia portion of Interstate 66 and the New Jersey Turnpike widening project through East Brunswick, New Jersey.

Area source models were developed in 1971 through 1974 by the ERT and ESL groups, but addressed a smaller fraction of total air pollution emissions, so that their use and need was not as widespread as the line source model, which enjoyed hundreds of different applications as early as the 1970s. Similarly photochemical models were developed primarily in the 1960s and 70s, but their use was more specialized and for regional needs, such as understanding smog formation in Los Angeles, California.

Soil Contamination

Excavation showing soil contamination at a disused gasworks.

Soil contamination or soil pollution as part of land degradation is caused by the presence of xenobiotic (human-made) chemicals or other alteration in the natural soil environment. It is typically caused by industrial activity, agricultural chemicals, or improper disposal of waste. The most common chemicals involved are petroleum hydrocarbons, polynuclear aromatic hydrocarbons (such as naphthalene and benzo(a)pyrene), solvents, pesticides, lead, and other heavy metals. Contamination is correlated with the degree of industrialization and intensity of chemical usage.

The concern over soil contamination stems primarily from health risks, from direct contact with the contaminated soil, vapors from the contaminants, and from secondary contamination of water supplies within and underlying the soil. Mapping of contaminated soil sites and the resulting cleanup are time consuming and expensive tasks, requiring extensive amounts of geology, hydrology, chemistry, computer modeling skills, and GIS in Environmental Contamination, as well as an appreciation of the history of industrial chemistry.

In North America and Western Europe the extent of contaminated land is best known, with many of countries in these areas having a legal framework to identify and deal with this environmental problem. Developing countries tend to be less tightly regulated despite some of them having undergone significant industrialization.

Causes

Soil pollution can be caused by the following (non-exhaustive list!):

- Oil drilling
- Mining and activities by other heavy industries
- Accidental spills as may happen during deforestation activities, etc.
- Corrosion of underground storage tanks (including piping used to transmit the contents)
- Acid rain (in turn caused by air pollution)
- Intensive farming
- Agrochemicals, such as pesticides, herbicides and fertilizers
- Industrial accidents
- Road debris
- Drainage of contaminated surface water into the soil
- Waste disposal
    o Oil and fuel dumping
    o Nuclear wastes
    o Direct discharge of industrial wastes to the soil
    o Landfill and illegal dumping
    o coal ash
    o Electronic waste
    o ammunitions and agents of war

The most common chemicals involved are petroleum hydrocarbons, solvents, pesticides, lead, and other heavy metals.

In a wider sense, genetically modified plants (GMP) can count as a risk factor for soils, because of their potential to affect the soil fauna. Any activity that leads to other forms of soil degradation (erosion, compaction, etc.) may indirectly worsen the contamination effects in that soil remediation becomes more tedious.

Historical deposition of coal ash used for residential, commercial, and industrial heating, as well as for industrial processes such as ore smelting, were a common source of contamination in areas that were industrialized before about 1960. Coal naturally concentrates lead and zinc during its formation, as well as other heavy metals to a lesser degree. When the coal is burned, most of these metals become concentrated in the ash (the principal exception being mercury). Coal ash and slag may contain sufficient lead to qualify as a "characteristic hazardous waste", defined in the USA as containing more than 5 mg/l of extractable lead using the TCLP procedure. In addition to lead, coal ash typically contains variable but significant concentrations of polynuclear aromatic hydrocarbons (PAHs; e.g., benzo(a)anthracene, benzo(b)fluoranthene, benzo(k)fluoranthene, benzo(a) pyrene, indeno(cd)pyrene, phenanthrene, anthracene, and others). These PAHs are known human carcinogens and the acceptable concentrations of them in soil are typically around 1 mg/kg. Coal ash and slag can be recognised by the presence of off-white grains in soil, gray heterogeneous soil, or (coal slag) bubbly, vesicular pebble-sized grains.

Treated sewage sludge, known in the industry as biosolids, has become controversial as a "fertilizer". As it is the byproduct of sewage treatment, it generally contains more contaminants such as organisms, pesticides, and heavy metals than other soil.

In the European Union, the Urban Waste Water Treatment Directive allows sewage sludge to be sprayed onto land. The volume is expected to double to 185,000 tons of dry solids in 2005. This has good agricultural properties due to the high nitrogen and phosphate content. In 1990/1991, 13% wet weight was sprayed onto 0.13% of the land; however, this is expected to rise 15 fold by 2005. Advocates say there is a need to control this so that pathogenic microorganisms do not get into water courses and to ensure that there is no accumulation of heavy metals in the top soil.

Pesticides and Herbicides

A pesticide is a substance or mixture of substances used to kill a pest. A pesticide may be a chemical substance, biological agent (such as a virus or bacteria), antimicrobial, disinfectant or device used against any pest. Pests include insects, plant pathogens, weeds, mollusks, birds, mammals, fish, nematodes (roundworms) and microbes that compete with humans for food, destroy property, spread or are a vector for disease or cause a nuisance. Although there are benefits to the use of pesticides, there are also drawbacks, such as potential toxicity to humans and other organisms.

Herbicides are used to kill weeds, especially on pavements and railways. They are similar to auxins and most are biodegradable by soil bacteria. However, one group derived from trinitrotoluene (2:4 D and 2:4:5 T) have the impurity dioxin, which is very toxic and causes fatality even in low concentrations. Another herbicide is Paraquat. It is highly toxic but it rapidly degrades in soil due to the action of bacteria and does not kill soil fauna.

Insecticides are used to rid farms of pests which damage crops. The insects damage not only standing crops but also stored ones and in the tropics it is reckoned that one third of the total production is lost during food storage. As with fungicides, the first insecticides used in the nineteenth century were inorganic e.g. Paris Green and other compounds of arsenic. Nicotine has also been used since the late eighteenth century.

There are now two main groups of synthetic insecticides -

1. Organochlorines include DDT, Aldrin, Dieldrin and BHC. They are cheap to produce, potent and persistent. DDT was used on a massive scale from the 1930s, with a peak of 72,000 tonnes used 1970. Then usage fell as the harmful environmental effects were realized. It was found worldwide in fish and birds and was even discovered in the snow in the Antarctic. It is only slightly soluble in water but is very soluble in the bloodstream. It affects the nervous and endocrine systems and causes the eggshells of birds to lack calcium causing them to be easily breakable. It is thought to be responsible for the decline of the numbers of birds of prey like ospreys and peregrine falcons in the 1950s - they are now recovering. As well as increased concentration via the food chain, it is known to enter via permeable membranes, so fish get it through their gills. As it has low water solubility, it tends to stay at the water surface, so organisms that live there are most affected. DDT found in fish that formed part of the human food chain caused concern, but the levels found in the liver, kidney and brain tissues was less than 1 ppm and in fat was 10 ppm, which was below the level likely to cause harm. However, DDT was banned in the UK and the United States to stop the further buildup of it in the food chain. U.S. manufacturers continued to sell DDT to developing countries, who could not afford the expensive replacement chemicals and who did not have such stringent regulations governing the use of pesticides.

2. Organophosphates, e.g. parathion, methyl parathion and about 40 other insecticides are available nationally. Parathion is highly toxic, methyl-parathion is less so and Malathion is generally considered safe as it has low toxicity and is rapidly broken down in the mammalian liver. This group works by preventing normal nerve transmission as cholinesterase is prevented from breaking down the transmitter substance acetylcholine, resulting in uncontrolled muscle movements.

Agents of War

The disposal of munitions, and a lack of care in manufacture of munitions caused by the urgency of production, can contaminate soil for extended periods. There is little published evidence on this type of contamination largely because of restrictions placed by Governments of many countries on the publication of material related to war effort. However, mustard gas stored during World War II has contaminated some sites for up to 50 years and the testing of Anthrax as a potential biological weapon contaminated the whole island of Gruinard

Health Effects

Contaminated or polluted soil directly affects human health through direct contact with soil or via inhalation of soil contaminants which have vaporized; potentially greater threats are posed by the infiltration of soil contamination into groundwater aquifers used for human consumption, sometimes in areas apparently far removed from any apparent source of above ground contamination.

Health consequences from exposure to soil contamination vary greatly depending on pollutant type, pathway of attack and vulnerability of the exposed population. Chronic exposure to chromium, lead and other metals, petroleum, solvents, and many pesticide and herbicide formulations can be carcinogenic, can cause congenital disorders, or can cause other chronic health conditions. Industrial or man-made concentrations of naturally occurring substances, such as nitrate and ammonia associated with livestock manure from agricultural operations, have also been identified as health hazards in soil and groundwater.

Chronic exposure to benzene at sufficient concentrations is known to be associated with higher incidence of leukemia. Mercury and cyclodienes are known to induce higher incidences of kidney damage and some irreversible diseases. PCBs and cyclodienes are linked to liver toxicity. Organo-phosphates and carbomates can induce a chain of responses leading to neuromuscular blockage. Many chlorinated solvents induce liver changes, kidney changes and depression of the central nervous system. There is an entire spectrum of further health effects such as headache, nausea, fatigue, eye irritation and skin rash for the above cited and other chemicals. At sufficient dosages a large number of soil contaminants can cause death by exposure via direct contact, inhalation or ingestion of contaminants in groundwater contaminated through soil.

The Scottish Government has commissioned the Institute of Occupational Medicine to undertake a review of methods to assess risk to human health from contaminated land. The overall aim of the project is to work up guidance that should be useful to Scottish Local Authorities in assessing whether sites represent a significant possibility of significant harm (SPOSH) to human health. It is envisaged that the output of the project will be a short document providing high level guidance on health risk assessment with reference to existing published guidance and methodologies that have been identified as being particularly relevant and helpful. The project will examine how policy guidelines have been developed for determining the acceptability of risks to human health and propose an approach for assessing what constitutes unacceptable risk in line with the criteria for SPOSH as defined in the legislation and the Scottish Statutory Guidance.

Ecosystem Effects

Not unexpectedly, soil contaminants can have significant deleterious consequences for ecosystems. There are radical soil chemistry changes which can arise from the presence of many hazardous chemicals even at low concentration of the contaminant species. These changes can manifest in the alteration of metabolism of endemic microorganisms and arthropods resident in a given soil envi-ronment. The result can be virtual eradication of some of the primary food chain, which in turn could have major consequences for predator or consumer species. Even if the chemical effect on lower life forms is small, the lower pyramid levels of the food chain may ingest alien chemicals, which normally become more concentrated for each consuming rung of the food chain. Many of these effects are now well known, such as the concentration of persistent DDT materials for avian consumers, leading to weakening of egg shells, increased chick mortality and potential extinction of species.

Effects occur to agricultural lands which have certain types of soil contamination. Contaminants typically alter plant metabolism, often causing a reduction in crop yields. This has a secondary ef-fect upon soil conservation, since the languishing crops cannot shield the Earth's soil from erosion. Some of these chemical contaminants have long half-lives and in other cases derivative chemicals are formed from decay of primary soil contaminants.

Cleanup Options

Cleanup or environmental remediation is analyzed by environmental scientists who utilize field measurement of soil chemicals and also apply computer models (GIS in Environmental

Contamination) for analyzing transport and fate of soil chemicals. Various technologies have been developed for remediation of oil-contaminated soil/ sediments There are several principal strategies for remediation:

- Excavate soil and take it to a disposal site away from ready pathways for human or sensitive ecosystem contact. This technique also applies to dredging of bay muds containing toxins.

- Aeration of soils at the contaminated site (with attendant risk of creating air pollution)

- Thermal remediation by introduction of heat to raise subsurface temperatures sufficiently high to volatize chemical contaminants out of the soil for vapour extraction. Technologies include ISTD, electrical resistance heating (ERH), and ET-DSP^{tm}.

- Bioremediation, involving microbial digestion of certain organic chemicals. Techniques used in bioremediation include landfarming, biostimulation and bioaugmentating soil biota with commercially available microflora.

- Extraction of groundwater or soil vapor with an active electromechanical system, with subsequent stripping of the contaminants from the extract.

- Containment of the soil contaminants (such as by capping or paving over in place).

- Phytoremediation, or using plants (such as willow) to extract heavy metals

- Mycoremediation, or using fungus to metabolize contaminants and accumulate heavy metals.

By Country

Various national standards for concentrations of particular contaminants include the United States EPA Region 9 Preliminary Remediation Goals (U.S. PRGs), the U.S. EPA Region 3 Risk Based Concentrations (U.S. EPA RBCs) and National Environment Protection Council of Australia Guideline on Investigation Levels in Soil and Groundwater.

People's Republic of China

The immense and sustained growth of the People's Republic of China since the 1970s has exacted a price from the land in increased soil pollution. The State Environmental Protection Administration believes it to be a threat to the environment, to food safety and to sustainable agriculture. According to a scientific sampling, 150 million mi (100,000 square kilometers) of China's cultivated land have been polluted, with contaminated water being used to irrigate a further 31.5 million mi (21,670 km^2.) and another 2 million mi (1,300 square kilometers) covered or destroyed by solid waste. In total, the area accounts for one-tenth of China's cultivatable land, and is mostly in economically developed areas. An estimated 12 million tonnes of grain are contaminated by heavy metals every year, causing direct losses of 20 billion yuan (US$2.57 billion).

United Kingdom

Generic guidance commonly used in the United Kingdom are the Soil Guideline Values published by the Department for Environment, Food and Rural Affairs (DEFRA) and the Environment

Agency. These are screening values that demonstrate the minimal acceptable level of a substance. Above this there can be no assurances in terms of significant risk of harm to human health. These have been derived using the Contaminated Land Exposure Assessment Model (CLEA UK). Certain input parameters such as Health Criteria Values, age and land use are fed into CLEA UK to obtain a probabilistic output.

Guidance by the Inter Departmental Committee for the Redevelopment of Contaminated Land (ICRCL) has been formally withdrawn by DEFRA, for use as a prescriptive document to determine the potential need for remediation or further assessment.

The CLEA model published by DEFRA and the Environment Agency (EA) in March 2002 sets a framework for the appropriate assessment of risks to human health from contaminated land, as required by Part IIA of the Environmental Protection Act 1990. As part of this framework, generic Soil Guideline Values (SGVs) have currently been derived for ten contaminants to be used as "intervention values". These values should not be considered as remedial targets but values above which further detailed assessment should be considered.

Three sets of CLEA SGVs have been produced for three different land uses, namely

- residential (with and without plant uptake)

- allotments

- commercial/industrial

It is intended that the SGVs replace the former ICRCL values. It should be noted that the CLEA SGVs relate to assessing chronic (long term) risks to human health and do not apply to the protection of ground workers during construction, or other potential receptors such as groundwater, buildings, plants or other ecosystems. The CLEA SGVs are not directly applicable to a site completely covered in hardstanding, as there is no direct exposure route to contaminated soils.

To date, the first ten of fifty-five contaminant SGVs have been published, for the following: arsenic, cadmium, chromium, lead, inorganic mercury, nickel, selenium ethyl benzene, phenol and toluene. Draft SGVs for benzene, naphthalene and xylene have been produced but their publication is on hold. Toxicological data (Tox) has been published for each of these contaminants as well as for benzo[a]pyrene, benzene, dioxins, furans and dioxin-like PCBs, naphthalene, vinyl chloride, 1,1,2,2 tetrachloroethane and 1,1,1,2 tetrachloroethane, 1,1,1 trichloroethane, tetrachloroethene, carbon tetrachloride, 1,2-dichloroethane, trichloroethene and xylene. The SGVs for ethyl benzene, phenol and toluene are dependent on the soil organic matter (SOM) content (which can be calculated from the total organic carbon (TOC) content). As an initial screen the SGVs for 1% SOM are considered to be appropriate.

Canada

India

In March 2009, the issue of Uranium poisoning in Punjab attracted press coverage. It was alleged to be caused by fly ash ponds of thermal power stations, which reportedly lead to severe birth defects in children in the Faridkot and Bhatinda districts of Punjab. The news reports claimed the uranium levels were more than 60 times the maximum safe limit. In 2012, the Government of India confirmed that the ground water in Malwa belt of Punjab has uranium metal that is 50% above the trace limits set by the United Nations' World Health Organization. Scientific studies, based on over 1000 samples from various sampling points, could not trace the source to fly ash and any sources from thermal power plants or industry as originally alleged. The study also revealed that the uranium concentration in ground water of Malwa district is not 60 times the WHO limits, but only 50% above the WHO limit in 3 locations. This highest concentration found in samples was less than those found naturally in ground waters currently used for human purposes elsewhere, such as Finland. Research is underway to identify natural or other sources for the uranium.

Water Pollution

Raw sewage and industrial waste in the New River as it passes from Mexicali to Calexico, California

Water pollution is the contamination of water bodies (e.g. lakes, rivers, oceans, aquifers and groundwater). This form of environmental degradation occurs when pollutants are directly or indirectly discharged into water bodies without adequate treatment to remove harmful compounds.

Water pollution affects the entire biosphere – plants and organisms living in these bodies of water. In almost all cases the effect is damaging not only to individual species and population, but also to the natural biological communities.

Introduction

Water pollution is a major global problem which requires ongoing evaluation and revision of water resource policy at all levels (international down to individual aquifers and wells). It has been suggested that water pollution is the leading worldwide cause of deaths and diseases, and that it accounts for the deaths of more than 14,000 people daily. An estimated 580 people in India die of water pollution related illness every day. About 90 percent of the water in the cities of China is polluted. As of 2007, half a billion Chinese had no access to safe drinking water. In addition to the acute problems of water pollution in developing countries, developed countries also continue to struggle with pollution problems. For example, in the most recent national report on water quality in the United States, 44 percent of assessed stream miles, 64 percent of assessed lake acres, and 30 percent of assessed bays and estuarine square miles were classified as polluted. The head of China's national development agency said in 2007 that one quarter the length of China's seven main rivers were so poisoned the water harmed the skin.

Water is typically referred to as polluted when it is impaired by anthropogenic contaminants and either does not support a human use, such as drinking water, or undergoes a marked shift in its ability to support its constituent biotic communities, such as fish. Natural phenomena such as volcanoes, algae blooms, storms, and earthquakes also cause major changes in water quality and the ecological status of water.

Categories

Although interrelated, surface water and groundwater have often been studied and managed as separate resources. Surface water seeps through the soil and becomes groundwater. Conversely, groundwater can also feed surface water sources. Sources of surface water pollution are generally grouped into two categories based on their origin.

Point Sources

Point source pollution – Shipyard – Rio de Janeiro.

Point source water pollution refers to contaminants that enter a waterway from a single, identifiable source, such as a pipe or ditch. Examples of sources in this category include discharges from a sewage treatment plant, a factory, or a city storm drain. The U.S. Clean Water Act (CWA) defines point source for regulatory enforcement purposes. The CWA definition of point source was amended in 1987 to include municipal storm sewer systems, as well as industrial storm water, such as from construction sites.

Non-point Sources

Nonpoint source pollution refers to diffuse contamination that does not originate from a single discrete source. NPS pollution is often the cumulative effect of small amounts of contaminants gathered from a large area. A common example is the leaching out of nitrogen compounds from fertilized agricultural lands. Nutrient runoff in storm water from "sheet flow" over an agricultural field or a forest are also cited as examples of NPS pollution.

Blue drain and yellow fish symbol used by the UK Environment Agency to raise awareness of the ecological impacts of contaminating surface drainage

Contaminated storm water washed off of parking lots, roads and highways, called urban runoff, is sometimes included under the category of NPS pollution. However, because this runoff is typically channeled into storm drain systems and discharged through pipes to local surface waters, it becomes a point source.

Groundwater Pollution

Interactions between groundwater and surface water are complex. Consequently, groundwater pollution, also referred to as groundwater contamination, is not as easily classified as surface water pollution. By its very nature, groundwater aquifers are susceptible to contamination from sources that may not directly affect surface water bodies, and the distinction of point vs. non-point source may be irrelevant. A spill or ongoing release of chemical or radionuclide contaminants into soil (located away from a surface water body) may not create point or non-point source pollution but can contaminate the aquifer below, creating a toxic plume. The movement of the plume, called a plume front, may be analyzed through a hydrological transport model or groundwater model. Analysis of groundwater contamination may focus on soil characteristics and site geology, hydrogeology, hydrology, and the nature of the contaminants.

Causes

The specific contaminants leading to pollution in water include a wide spectrum of chemicals, pathogens, and physical changes such as elevated temperature and discoloration. While many of the chemicals and substances that are regulated may be naturally occurring (calcium, sodium, iron, manganese, etc.) the concentration is often the key in determining what is a natural component of water and what is a contaminant. High concentrations of naturally occurring substances can have negative impacts on aquatic flora and fauna.

Oxygen-depleting substances may be natural materials such as plant matter (e.g. leaves and grass) as well as man-made chemicals. Other natural and anthropogenic substances may cause turbidity (cloudiness) which blocks light and disrupts plant growth, and clogs the gills of some fish species.

Many of the chemical substances are toxic. Pathogens can produce waterborne diseases in either human or animal hosts. Alteration of water's physical chemistry includes acidity (change in pH), electrical conductivity, temperature, and eutrophication. Eutrophication is an increase in the concentration of chemical nutrients in an ecosystem to an extent that increases in the primary productivity of the ecosystem. Depending on the degree of eutrophication, subsequent negative environmental effects such as anoxia (oxygen depletion) and severe reductions in water quality may occur, affecting fish and other animal populations.

Pathogens

Disease-causing microorganisms are referred to as pathogens. Although the vast majority of bacteria are either harmless or beneficial, a few pathogenic bacteria can cause disease. Coliform bacteria, which are not an actual cause of disease, are commonly used as a bacterial indicator of water pollution. Other microorganisms sometimes found in surface waters that have caused human health problems include:

- *Burkholderia pseudomallei*

- *Cryptosporidium parvum*

- *Giardia lamblia*

- *Salmonella*

- *Norovirus* and other viruses

- *Parasitic worms including the Schistosoma type*

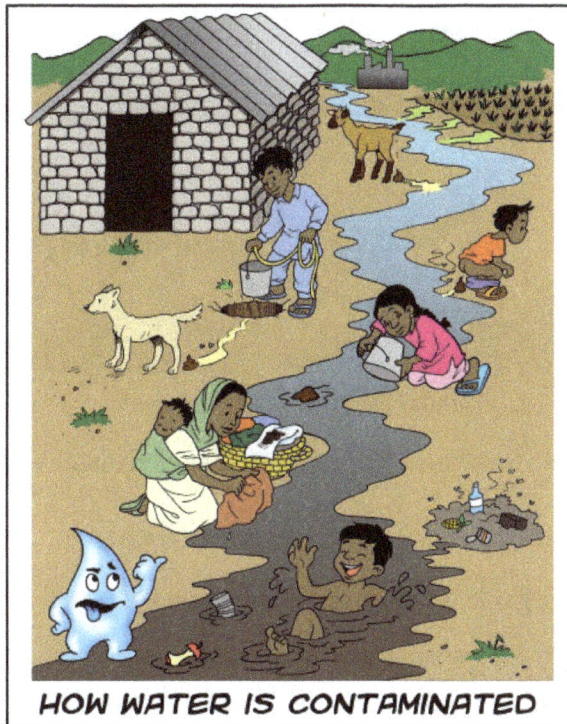

Poster to teach people in South Asia about human activities leading to the pollution of water sources

A manhole cover unable to contain a sanitary sewer overflow.

High levels of pathogens may result from on-site sanitation systems (septic tanks, pit latrines) or inadequately treated sewage discharges. This can be caused by a sewage plant designed with less than secondary treatment (more typical in less-developed countries). In developed countries, older cities with aging infrastructure may have leaky sewage collection systems (pipes, pumps, valves), which can cause sanitary sewer overflows. Some cities also have combined sewers, which may discharge untreated sewage during rain storms.

Muddy river polluted by sediment.

Pathogen discharges may also be caused by poorly managed livestock operations.

Organic, Inorganic and Macroscopic Contaminants

Contaminants may include organic and inorganic substances.

A garbage collection boom in an urban-area stream in Auckland, New Zealand.

Organic water pollutants include:

- Detergents

- Disinfection by-products found in chemically disinfected drinking water, such as chloroform

- Food processing waste, which can include oxygen-demanding substances, fats and grease

- Insecticides and herbicides, a huge range of organohalides and other chemical compounds

- Petroleum hydrocarbons, including fuels (gasoline, diesel fuel, jet fuels, and fuel oil) and lubricants (motor oil), and fuel combustion byproducts, from storm water runoff

- Volatile organic compounds, such as industrial solvents, from improper storage.

- Chlorinated solvents, which are dense non-aqueous phase liquids, may fall to the bottom of reservoirs, since they don't mix well with water and are denser.

    o Polychlorinated biphenyl (PCBs)

    o Trichloroethylene

- Perchlorate

- Various chemical compounds found in personal hygiene and cosmetic products

- Drug pollution involving pharmaceutical drugs and their metabolites

Inorganic water pollutants include:

- Acidity caused by industrial discharges (especially sulfur dioxide from power plants)

- Ammonia from food processing waste

- Chemical waste as industrial by-products

- Fertilizers containing nutrients--nitrates and phosphates—which are found in storm water runoff from agriculture, as well as commercial and residential use

- Heavy metals from motor vehicles (via urban storm water runoff) and acid mine drainage

- Silt (sediment) in runoff from construction sites, logging, slash and burn practices or land clearing sites.

Macroscopic pollution – large visible items polluting the water – may be termed "floatables" in an urban storm water context, or marine debris when found on the open seas, and can include such items as:

- Trash or garbage (e.g. paper, plastic, or food waste) discarded by people on the ground, along with accidental or intentional dumping of rubbish, that are washed by rainfall into storm drains and eventually discharged into surface waters

- Nurdles, small ubiquitous waterborne plastic pellets

- Shipwrecks, large derelict ships.

The Brayton Point Power Station in Massachusetts discharges heated water to Mount Hope Bay.

Thermal Pollution

Thermal pollution is the rise or fall in the temperature of a natural body of water caused by human influence. Thermal pollution, unlike chemical pollution, results in a change in the physical properties of water. A common cause of thermal pollution is the use of water as a coolant by power plants and industrial manufacturers. Elevated water temperatures decrease oxygen levels, which can kill fish and alter food chain composition, reduce species biodiversity, and foster invasion by new thermophilic species. Urban runoff may also elevate temperature in surface waters.

Thermal pollution can also be caused by the release of very cold water from the base of reservoirs into warmer rivers.

Transport and Chemical Reactions of Water Pollutants

Most water pollutants are eventually carried by rivers into the oceans. In some areas of the world the influence can be traced one hundred miles from the mouth by studies using hydrology transport models. Advanced computer models such as SWMM or the DSSAM Model have been used in many locations worldwide to examine the fate of pollutants in aquatic systems. Indicator filter feeding species such as copepods have also been used to study pollutant fates in the New York Bight, for example. The highest toxin loads are not directly at the mouth of the Hudson River, but 100 km (62 mi) south, since several days are required for incorporation into planktonic tissue. The Hudson discharge flows south along the coast due to the coriolis force. Further south are areas of oxygen depletion caused by chemicals using up oxygen and by algae blooms, caused by excess nutrients from algal cell death and decomposition. Fish and shellfish kills have been reported, because toxins climb the food chain after small fish consume copepods, then large fish eat smaller fish, etc. Each successive step up the food chain causes a cumulative concentration of pollutants such as heavy metals (e.g. mercury) and persistent organic pollutants such as DDT. This is known as bio-magnification, which is occasionally used interchangeably with bio-accumulation.

Large gyres (vortexes) in the oceans trap floating plastic debris. The North Pacific Gyre, for example, has collected the so-called "Great Pacific Garbage Patch", which is now estimated to be one hundred times the size of Texas. Plastic debris can absorb toxic chemicals from ocean pollution, potentially poisoning any creature that eats it. Many of these long-lasting pieces wind up in the stomachs of marine birds and animals. This results in obstruction of digestive pathways, which leads to reduced appetite or even starvation.

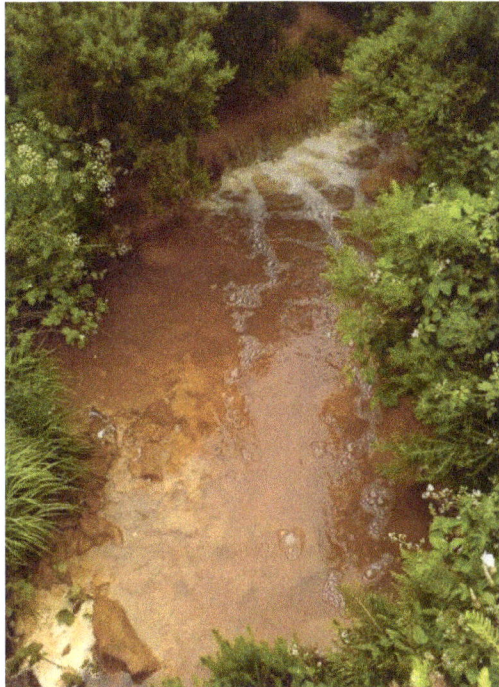

A polluted river draining an abandoned copper mine on Anglesey

Many chemicals undergo reactive decay or chemical change, especially over long periods of time in groundwater reservoirs. A noteworthy class of such chemicals is the chlorinated hydrocarbons such as trichloroethylene (used in industrial metal degreasing and electronics manufacturing) and tetrachloroethylene used in the dry cleaning industry. Both of these chemicals, which are carcinogens themselves, undergo partial decomposition reactions, leading to new hazardous chemicals (including dichloroethylene and vinyl chloride).

Groundwater pollution is much more difficult to abate than surface pollution because groundwater can move great distances through unseen aquifers. Non-porous aquifers such as clays partially purify water of bacteria by simple filtration (adsorption and absorption), dilution, and, in some cases, chemical reactions and biological activity; however, in some cases, the pollutants merely transform to soil contaminants. Groundwater that moves through open fractures and caverns is not filtered and can be transported as easily as surface water. In fact, this can be aggravated by the human tendency to use natural sinkholes as dumps in areas of karst topography.

There are a variety of secondary effects stemming not from the original pollutant, but a derivative condition. An example is silt-bearing surface runoff, which can inhibit the penetration of sunlight through the water column, hampering photosynthesis in aquatic plants.

Measurement

Water pollution may be analyzed through several broad categories of methods: physical, chemical and biological. Most involve collection of samples, followed by specialized analytical tests. Some methods may be conducted *in situ*, without sampling, such as temperature. Government agencies and research organizations have published standardized, validated analytical test methods to facilitate the comparability of results from disparate testing events.

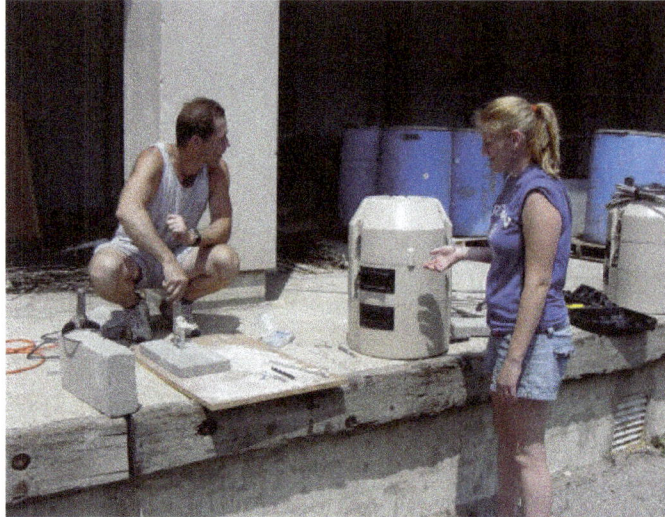

Environmental scientists preparing water autosamplers.

Sampling

Sampling of water for physical or chemical testing can be done by several methods, depending on the accuracy needed and the characteristics of the contaminant. Many contamination events are sharply restricted in time, most commonly in association with rain events. For this reason "grab" samples are often inadequate for fully quantifying contaminant levels. Scientists gathering this type of data often employ auto-sampler devices that pump increments of water at either time or discharge intervals.

Sampling for biological testing involves collection of plants and/or animals from the surface water body. Depending on the type of assessment, the organisms may be identified for biosurveys (population counts) and returned to the water body, or they may be dissected for bioassays to determine toxicity.

Physical Testing

Common physical tests of water include temperature, solids concentrations (e.g., total suspended solids (TSS)) and turbidity.

Chemical Testing

Water samples may be examined using the principles of analytical chemistry. Many published test methods are available for both organic and inorganic compounds. Frequently used methods include pH, biochemical oxygen demand (BOD), chemical oxygen demand (COD), nutrients (nitrate and phosphorus compounds), metals (including copper, zinc, cadmium, lead and mercury), oil and grease, total petroleum hydrocarbons (TPH), and pesticides.

Biological Testing

Biological testing involves the use of plant, animal, and/or microbial indicators to monitor the health of an aquatic ecosystem. They are any biological species or group of species whose function, population, or status can reveal what degree of ecosystem or environmental integrity is present. One example of a group of bio-indicators are the copepods and other small water crustaceans that are present in many water bodies. Such organisms can be monitored for changes (biochemical, physiological, or behavioral) that may indicate a problem within their ecosystem.

Control of Pollution

Decisions on the type and degree of treatment and control of wastes, and the disposal and use of adequately treated wastewater, must be based on a consideration all the technical factors of each drainage basin, in order to prevent any further contamination or harm to the environment.

Sewage Treatment

Deer Island Wastewater Treatment Plant serving Boston, Massachusetts and vicinity.

In urban areas of developed countries, domestic sewage is typically treated by centralized sewage treatment plants. Well-designed and operated systems (i.e., secondary treatment or better) can remove 90 percent or more of the pollutant load in sewage. Some plants have additional systems to remove nutrients and pathogens.

Cities with sanitary sewer overflows or combined sewer overflows employ one or more engineering approaches to reduce discharges of untreated sewage, including:

- utilizing a green infrastructure approach to improve storm water management capacity throughout the system, and reduce the hydraulic overloading of the treatment plant

- repair and replacement of leaking and malfunctioning equipment

- increasing overall hydraulic capacity of the sewage collection system (often a very expensive option).

A household or business not served by a municipal treatment plant may have an individual septic tank, which pre-treats the wastewater on site and infiltrates it into the soil.

Industrial Wastewater Treatment

Dissolved air flotation system for treating industrial wastewater.

Some industrial facilities generate ordinary domestic sewage that can be treated by municipal facilities. Industries that generate wastewater with high concentrations of conventional pollutants (e.g. oil and grease), toxic pollutants (e.g. heavy metals, volatile organic compounds) or other non-conventional pollutants such as ammonia, need specialized treatment systems. Some of these facilities can install a pre-treatment system to remove the toxic components, and then send the partially treated wastewater to the municipal system. Industries generating large volumes of wastewater typically operate their own complete on-site treatment systems. Some industries have been successful at redesigning their manufacturing processes to reduce or eliminate pollutants, through a process called pollution prevention.

Heated water generated by power plants or manufacturing plants may be controlled with:

- cooling ponds, man-made bodies of water designed for cooling by evaporation, convection, and radiation

- cooling towers, which transfer waste heat to the atmosphere through evaporation and/or heat transfer

- cogeneration, a process where waste heat is recycled for domestic and/or industrial heating purposes.

Riparian buffer lining a creek in Iowa.

Agricultural Wastewater Treatment

Non point source controls Sediment (loose soil) washed off fields is the largest source of agricultural pollution in the United States. Farmers may utilize erosion controls to reduce runoff flows and retain soil on their fields. Common techniques include contour plowing, crop mulching, crop rotation, planting perennial crops and installing riparian buffers.

Nutrients (nitrogen and phosphorus) are typically applied to farmland as commercial fertilizer, animal manure, or spraying of municipal or industrial wastewater (effluent) or sludge. Nutrients may also enter runoff from crop residues, irrigation water, wildlife, and atmospheric deposition. Farmers can develop and implement nutrient management plans to reduce excess application of nutrients and reduce the potential for nutrient pollution.

To minimize pesticide impacts, farmers may use Integrated Pest Management (IPM) techniques (which can include biological pest control) to maintain control over pests, reduce reliance on chemical pesticides, and protect water quality.

Feedlot in the United States

Point source wastewater treatment Farms with large livestock and poultry operations, such as factory farms, are called *concentrated animal feeding operations* or *feedlots* in the US and are being subject to increasing government regulation. Animal slurries are usually treated by containment in anaerobic lagoons before disposal by spray or trickle application to grassland. Constructed wetlands are sometimes used to facilitate treatment of animal wastes. Some animal slurries are treated by mixing with straw and composted at high temperature to produce a bacteriologically sterile and friable manure for soil improvement.

Erosion and Sediment Control From Construction Sites

Sediment from construction sites is managed by installation of:

- erosion controls, such as mulching and hydroseeding, and
- sediment controls, such as sediment basins and silt fences.

Silt fence installed on a construction site.

Discharge of toxic chemicals such as motor fuels and concrete washout is prevented by use of:

- spill prevention and control plans, and

- specially designed containers (e.g. for concrete washout) and structures such as overflow controls and diversion berms.

Control of Urban Runoff (Storm Water)

Retention basin for controlling urban runoff

Effective control of urban runoff involves reducing the velocity and flow of storm water, as well as reducing pollutant discharges. Local governments use a variety of storm water management techniques to reduce the effects of urban runoff. These techniques, called best management practices (BMPs) in the U.S., may focus on water quantity control, while others focus on improving water quality, and some perform both functions.

Pollution prevention practices include low-impact development techniques, installation of green roofs and improved chemical handling (e.g. management of motor fuels & oil, fertilizers and pesticides). Runoff mitigation systems include infiltration basins, bioretention systems, constructed wetlands, retention basins and similar devices.

Thermal pollution from runoff can be controlled by storm water management facilities that absorb the runoff or direct it into groundwater, such as bioretention systems and infiltration basins. Retention basins tend to be less effective at reducing temperature, as the water may be heated by the sun before being discharged to a receiving stream.

Noise Pollution

Traffic is the main source of noise pollution in cities.

A Qantas Airways Boeing 747-400 passes close to houses shortly before landing at London Heathrow Airport.

Noise pollution or noise disturbance is the disturbing or excessive noise that may harm the activity or balance of human or animal life. The source of most outdoor noise worldwide is mainly caused by machines and transportation systems, motor vehicles, aircraft, and trains. Outdoor noise is summarized by the word environmental noise. Poor urban planning may give rise to noise pollution, since side-by-side industrial and residential buildings can result in noise pollution in the residential areas. Documented problems associated with urban noise go back as far as Ancient Rome.

Outdoor noise can be caused by machines, construction activities, and music performances, especially in some workplaces. Noise-induced hearing loss can be caused by outside (e.g. trains) or inside (e.g. music) noise.

High noise levels can contribute to cardiovascular effects in humans and an increased incidence of coronary artery disease. In animals, noise can increase the risk of death by altering predator or prey detection and avoidance, interfere with reproduction and navigation, and contribute to permanent hearing loss.

Health

Humans

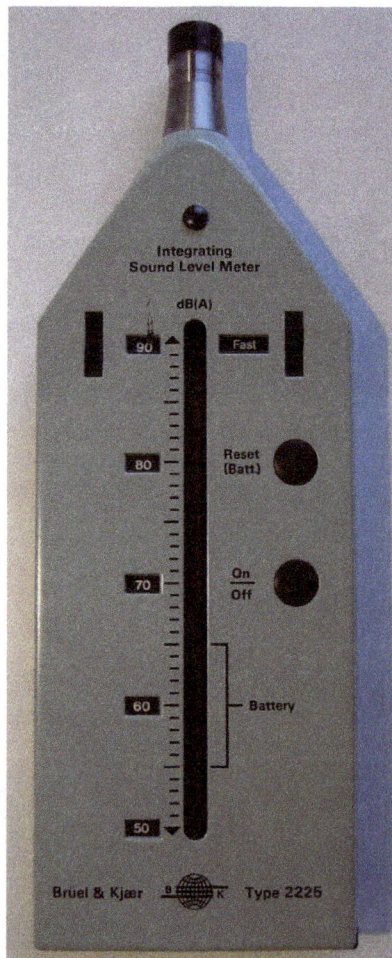

A sound level meter, a basic tool in measuring sound.

Noise pollution affects both health and behavior. Unwanted sound (noise) can damage psychological health. Noise pollution can cause hypertension, high stress levels, tinnitus, hearing loss, sleep disturbances, and other harmful effects.

Sound becomes unwanted when it either interferes with normal activities such as sleeping, conversation, or disrupts or diminishes one's quality of life.

Reaction to noise

Chronic exposure to noise may cause noise-induced hearing loss. Older males exposed to significant occupational noise demonstrate more significantly reduced hearing sensitivity than their non-exposed peers, though differences in hearing sensitivity decrease with time and the two groups are indistinguishable by age 79. A comparison of Maaban tribesmen, who were insignificantly exposed to transportation or industrial noise, to a typical U.S. population showed that chronic exposure to moderately high levels of environmental noise contributes to hearing loss.

High noise levels can result in cardiovascular effects and exposure to moderately high levels during a single eight-hour period causes a statistical rise in blood pressure of five to ten points and an increase in stress, and vasoconstriction leading to the increased blood pressure noted above, as well as to increased incidence of coronary artery disease.

Wildlife

Noise can have a detrimental effect on wild animals, increasing the risk of death by changing the delicate balance in predator or prey detection and avoidance, and interfering the use of the sounds in communication, especially in relation to reproduction and in navigation. Acoustic overexposure can lead to temporary or permanent loss of hearing.

An impact of noise on wild animal life is the reduction of usable habitat that noisy areas may cause, which in the case of endangered species may be part of the path to extinction. Noise pollution may have caused the death of certain species of whales that beached themselves after being exposed to the loud sound of military sonar.

Noise also makes species communicate more loudly, which is called Lombard vocal response. Scientists and researchers have conducted experiments that show whales' song length is longer when submarine-detectors are on. If creatures do not "speak" loudly enough, their voice will be masked by anthropogenic sounds. These unheard voices might be warnings, finding of prey, or preparations of net-bubbling. When one species begins speaking more loudly, it will mask other species' voice, causing the whole ecosystem eventually to speak more loudly.

Marine invertebrates, such as crabs (*Carcinus maenas*), have also been shown to be impacted by ship noise. Larger crabs were noted to be impacted more by the sounds than smaller crabs. Repeated exposure to the sounds did lead to acclimatization.

European robins living in urban environments are more likely to sing at night in places with high levels of noise pollution during the day, suggesting that they sing at night because it is quieter, and their message can propagate through the environment more clearly. The same study showed that daytime noise was a stronger predictor of nocturnal singing than night-time light pollution, to which the phenomenon often is attributed.

Zebra finches become less faithful to their partners when exposed to traffic noise. This could alter a population's evolutionary trajectory by selecting traits, sapping resources normally devoted to other activities and thus leading to profound genetic and evolutionary consequences.

Noise Mitigation

The sound tube in Melbourne, Australia is designed to reduce roadway noise without detracting from the area's aesthetics

A man wears ear muffs for protection against noise pollution, 1973.

Roadway noise can be reduced by the use of noise barriers, limitation of vehicle speeds, alteration of roadway surface texture, limitation of heavy vehicles, use of traffic controls that smooth vehicle flow to reduce braking and acceleration, and tire design. An important factor in applying these strategies is a computer model for roadway noise, that is capable of addressing local topography, meteorology, traffic operations, and hypothetical mitigation. Costs of building-in mitigation can be modest, provided these solutions are sought in the planning stage of a roadway project.

Aircraft noise can be reduced by using quieter jet engines. Altering flight paths and time of day runway has benefitted residents near airports.

Industrial noise has been addressed since the 1930s via redesign of industrial equipment, shock mounted assemblies and physical barriers in the workplace. In recent years, Buy Quiet programs and initiatives have arisen in an effort to combat occupational noise exposures. These programs promote the purchase of quieter tools and equipment and encourage manufacturers to design quieter equipment. The National Institute for Occupational Health has created a database of industrial equipment with decibel levels noted.

Legal Status

Up until the 1970s governments tended to view noise as a "nuisance" rather than an environmental problem.

Many conflicts over noise pollution are handled by negotiation between the emitter and the receiver. Escalation procedures vary by country, and may include action in conjunction with local authorities, in particular the police.

India

Noise pollution is a major problem in India. The government of India has rules & regulations against firecrackers and loudspeakers, but enforcement is extremely lax. Awaaz (sound) Foundation is an Indian NGO working to control noise pollution from various sources through advocacy, public interest litigation, awareness, and educational campaigns since 2003.

United Kingdom

Figures compiled by rockwool, the mineral wool insulation manufacturer, based on responses from local authorities to a Freedom of Information Act (FOI) request reveal in the period April 2008 – 2009 UK councils received 315,838 complaints about noise pollution from private residences. This resulted in environmental health officers across the UK serving 8,069 noise abatement notices or citations under the terms of the Anti-Social Behaviour (Scotland) Act. In the last 12 months, 524 confiscations of equipment have been authorized involving the removal of powerful speakers, stereos and televisions. Westminster City Council has received more complaints per head of population than any other district in the UK with 9,814 grievances about noise, which equates to 42.32 complaints per thousand residents. Eight of the top 10 councils ranked by complaints per 1,000 residents are located in London.

United States

There are federal standards for highway and aircraft noise; states and local governments typically have very specific statutes on building codes, urban planning, and roadway development.

Noise laws and ordinances vary widely among municipalities and indeed do not even exist in some cities. An ordinance may contain a general prohibition against making noise that is a nuisance, or it may set out specific guidelines for the level of noise allowable at certain times of the day and for certain activities.

The Environmental Protection Agency retains authority to investigate and study noise and its effect, disseminate information to the public regarding noise pollution and its adverse health effects, respond to inquiries on matters related to noise, and evaluate the effectiveness of existing regulations for protecting the public health and welfare, pursuant to the Noise Control Act of 1972 and the Quiet Communities Act of 1978.

New York City instituted the first comprehensive noise code in 1985. The Portland Noise Code includes potential fines of up to $5000 per infraction and is the basis for other major U.S. and Canadian city noise ordinances.

References

- Davis, Devra (2002). When Smoke Ran Like Water: Tales of Environmental Deception and the Battle Against Pollution. Basic Books. ISBN 0-465-01521-2.

- Turner, D.B. (1994). Workbook of atmospheric dispersion estimates: an introduction to dispersion modeling (2nd ed.). CRC Press. ISBN 1-56670-023-X.

- G. Allen Burton, Jr., Robert Pitt (2001). Stormwater Effects Handbook: A Toolbox for Watershed Managers, Scientists, and Engineers. New York: CRC/Lewis Publishers. ISBN 0-87371-924-7.

- Goel, P.K. (2006). Water Pollution - Causes, Effects and Control. New Delhi: New Age International. p. 179. ISBN 978-81-224-1839-2.

- Kennish, Michael J. (1992). Ecology of Estuaries: Anthropogenic Effects. Marine Science Series. Boca Raton, FL: CRC Press. pp. 415–17. ISBN 978-0-8493-8041-9.

- Laws, Edward A. (2000). Aquatic Pollution: An Introductory Text. New York: John Wiley and Sons. p. 430. ISBN 978-0-471-34875-7.

- George, Rebecca; Joy, Varsha; S, Aiswarya; Jacob, Priya A. "Treatment Methods for Contaminated Soils - Translating Science into Practice" (PDF). International Journal of Education and Applied Research. Retrieved February 19, 2016.

- St. Fleur, Nicholas (10 November 2015). "Atmospheric Greenhouse Gas Levels Hit Record, Report Says". New York Times. Retrieved 11 November 2015.

- Ritter, Karl (9 November 2015). "UK: In 1st, global temps average could be 1 degree C higher". AP News. Retrieved 11 November 2015.

- Cole, Steve; Gray, Ellen (14 December 2015). "New NASA Satellite Maps Show Human Fingerprint on Global Air Quality". NASA. Retrieved 14 December 2015.

- "Bucknell tent death: Hannah Thomas-Jones died from carbon monoxide poisoning". BBC News. 17 January 2013. Retrieved 22 September 2015.

- Gallagher, James (17 December 2015). "Cancer is not just 'bad luck' but down to environment, study suggests". BBC. Retrieved 17 December 2015.

- European Court of Justice, CURIA (2008). "PRESS RELEASE No 58/08 Judgment of the Court of Justice in Case C-237/07" (PDF). Retrieved 24 January 2015.

- House of Commons Environmental Audit Committee (2010). "Environmental Audit Committee - Fifth Report Air Quality". Retrieved 24 January 2015.

- BInnes, Emma (6 June 2014). "Air pollution 'can cause changes in the brain seen in autism and schizophrenia'". Daily Mail. Retrieved 8 June 2014.

- "New Evidence Links Air Pollution to Autism, Schizophrenia". University of Rochester Medical Center. 6 June 2014. Retrieved 8 June 2014.

Effects of Pollution on Environment

Contamination of nature through human activity is the most dangerous threat to our natural resources. A lack of scientific and systematic pollution management leads to larger number of areas becoming polluted by contamination and waste accumulation. The effects of pollution are best understood in confluence with the major topics listed in the following chapter.

Global Warming

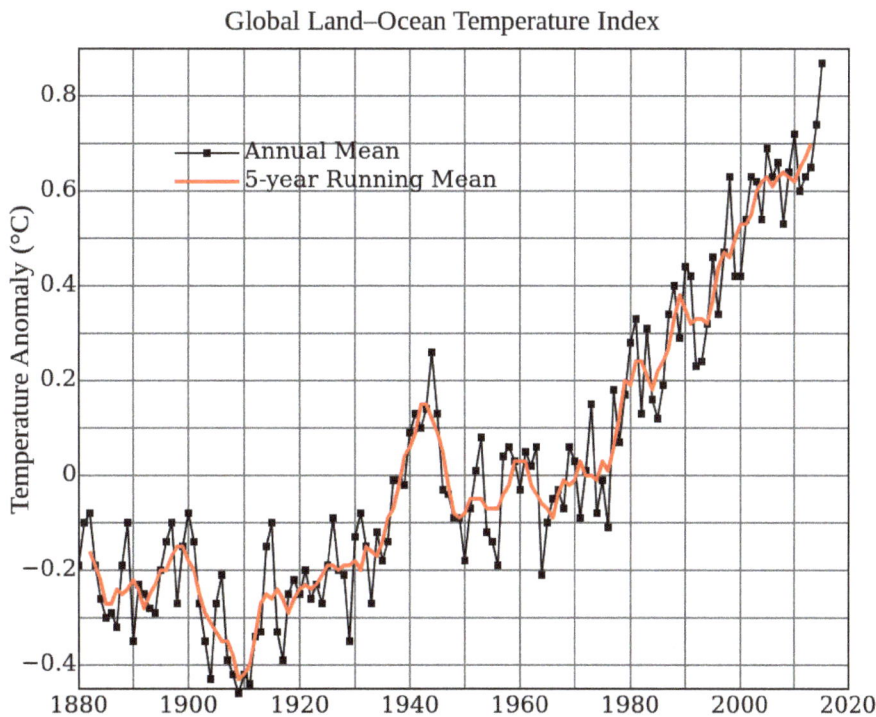

Global Land–Ocean Temperature Index

Global mean surface temperature change from 1880 to 2015, relative to the 1951–1980 mean. The black line is the annual mean and the red line is the 5-year running mean. Source: NASA GISS.

Global warming and climate change are terms for the observed century-scale rise in the average temperature of the Earth's climate system and its related effects. Multiple lines of scientific evidence show that the climate system is warming. Although the increase of near-surface atmospheric temperature is the measure of global warming often reported in the popular press, most of the additional energy stored in the climate system since 1970 has gone into ocean warming. The remainder has melted ice and warmed the continents and atmosphere. Many of the observed changes since the 1950s are unprecedented over tens to thousands of years.

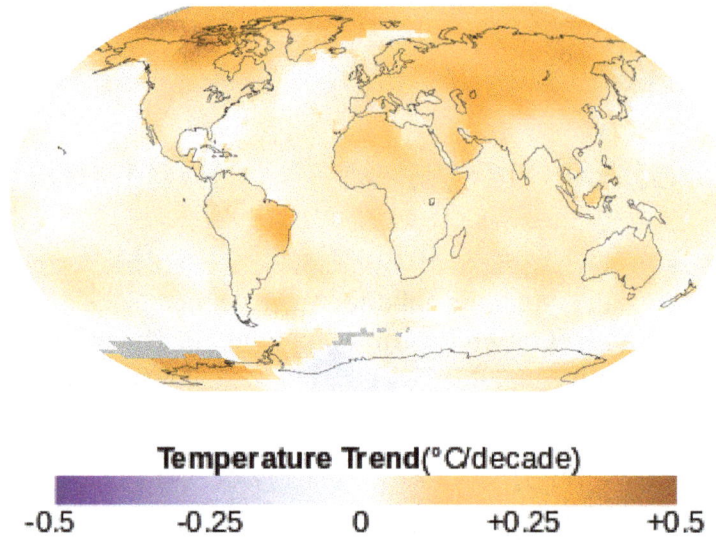

World map showing surface temperature trends (°C per decade) between 1950 and 2014. Source: NASA GISS.

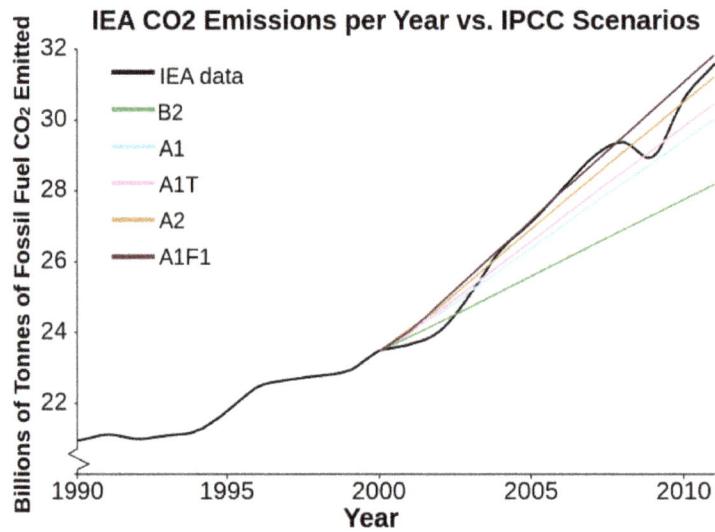

Fossil fuel related carbon dioxide (CO_2) emissions compared to five of the IPCC's "SRES" emissions scenarios, published in 2000. The dips are related to global recessions. Image source: Skeptical Science.

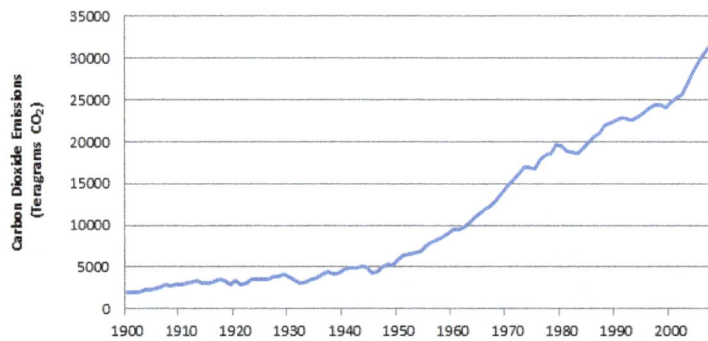

Fossil fuel related carbon dioxide emissions over the 20th century. Image source: EPA.

Scientific understanding of global warming is increasing. The Intergovernmental Panel on Climate Change (IPCC) reported in 2014 that scientists were more than 95% certain that global warming is mostly being caused by human (anthropogenic) activities, mainly increasing concentrations of greenhouse gases such as carbon dioxide (CO_2). Human-made carbon dioxide continues to increase above levels not seen in hundreds of thousands of years: currently, about half of the carbon dioxide released from the burning of fossil fuels is not absorbed by vegetation and the oceans and remains in the atmosphere. Climate model projections summarized in the report indicated that during the 21st century the global surface temperature is likely to rise a further 0.3 to 1.7 °C (0.5 to 3.1 °F) for their lowest emissions scenario using stringent mitigation and 2.6 to 4.8 °C (4.7 to 8.6 °F) for their highest. These findings have been recognized by the national science academies of the major industrialized nations and are not disputed by any scientific body of national or international standing.

Future climate change and associated impacts will differ from region to region around the globe. Anticipated effects include warming global temperature, rising sea levels, changing precipitation, and expansion of deserts in the subtropics. Warming is expected to be greater over land than over the oceans and greatest in the Arctic, with the continuing retreat of glaciers, permafrost and sea ice. Other likely changes include more frequent extreme weather events including heat waves, droughts, heavy rainfall with floods and heavy snowfall; ocean acidification; and species extinctions due to shifting temperature regimes. Effects significant to humans include the threat to food security from decreasing crop yields and the abandonment of populated areas due to rising sea levels. Because the climate system has a large "inertia" and CO_2 will stay in the atmosphere for a long time, many of these effects will not only exist for decades or centuries, but will persist for tens of thousands of years.

Possible societal responses to global warming include mitigation by emissions reduction, adaptation to its effects, building systems resilient to its effects, and possible future climate engineering. Most countries are parties to the United Nations Framework Convention on Climate Change (UNFCCC), whose ultimate objective is to prevent dangerous anthropogenic climate change. The UNFCCC have adopted a range of policies designed to reduce greenhouse gas emissions and to assist in adaptation to global warming. Parties to the UNFCCC had agreed that deep cuts in emissions are required and as first target the future global warming should be limited to below 2.0 °C (3.6 °F) relative to the pre-industrial level, while the Paris Agreement of 2015 stated that the parties will also "pursue efforts to" limit the temperature increase to 1.5 °F (0.8 °C).

Public reactions to global warming and general fears of its effects are also steadily on the rise, with a global 2015 Pew Research Center report showing a median of 54% who consider it "a very serious problem". There are, however, significant regional differences. Notably, Americans and Chinese, whose economies are responsible for the greatest annual CO2 emissions, are among the least concerned.

Observed Temperature Changes

The global average (land and ocean) surface temperature shows a warming of 0.85 [0.65 to 1.06] °C in the period 1880 to 2012, based on multiple independently produced datasets. Earth's average surface temperature rose by 0.74±0.18 °C over the period 1906–2005. The rate of warming almost doubled for the last half of that period (0.13±0.03 °C per decade, versus 0.07±0.02 °C per decade).

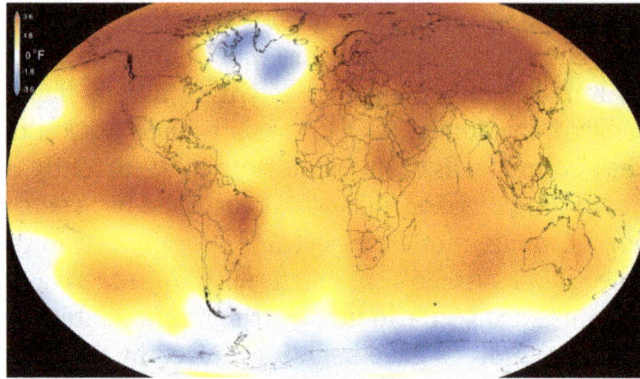

2015 – Warmest Global Year on Record (since 1880) – Colors indicate temperature anomalies (NASA/NOAA; 20 January 2016).

Energy change inventory, 1971-2010

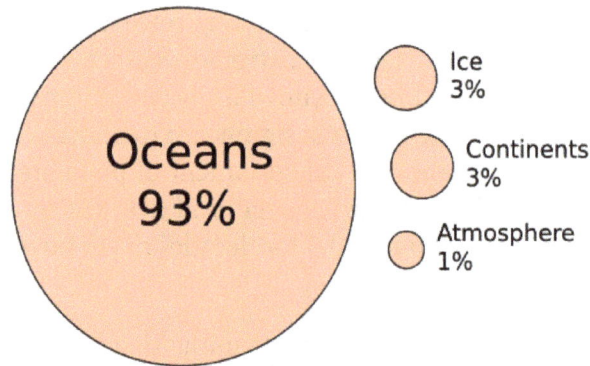

Earth has been in *radiative imbalance* since at least the 1970s, where less energy leaves the atmosphere than enters it. Most of this extra energy has been absorbed by the oceans. It is very likely that human activities substantially contributed to this increase in ocean heat content.

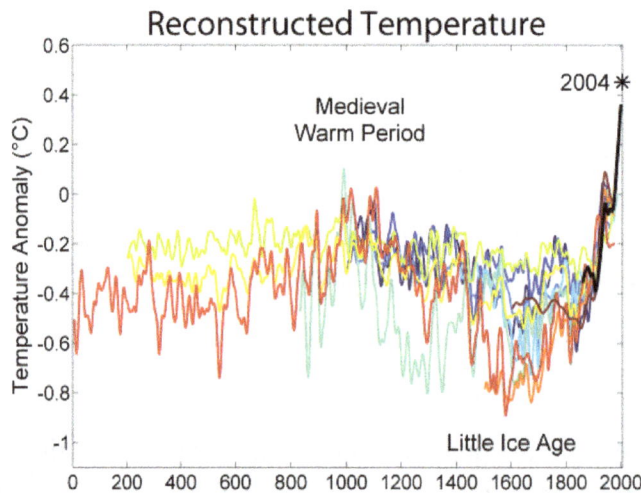

Two millennia of mean surface temperatures according to different reconstructions from climate proxies, each smoothed on a decadal scale, with the instrumental temperature record overlaid in black.

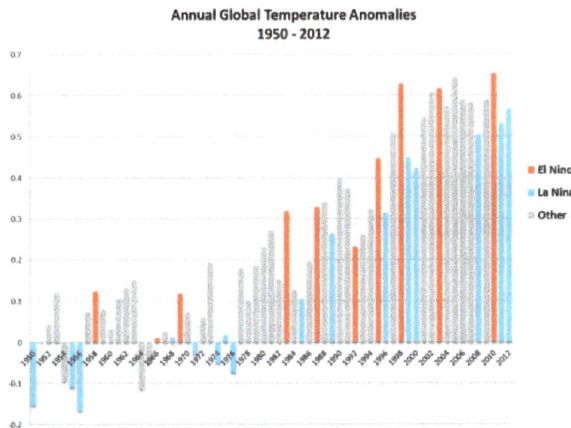

NOAA graph of Global Annual Temperature Anomalies 1950–2012, showing the El Niño Southern Oscillation

The average temperature of the lower troposphere has increased between 0.13 and 0.22 °C (0.23 and 0.40 °F) per decade since 1979, according to satellite temperature measurements. Climate proxies show the temperature to have been relatively stable over the one or two thousand years before 1850, with regionally varying fluctuations such as the Medieval Warm Period and the Little Ice Age.

The warming that is evident in the instrumental temperature record is consistent with a wide range of observations, as documented by many independent scientific groups. Examples include sea level rise, widespread melting of snow and land ice, increased heat content of the oceans, increased humidity, and the earlier timing of spring events, e.g., the flowering of plants. The probability that these changes could have occurred by chance is virtually zero.

Trends

Temperature changes vary over the globe. Since 1979, land temperatures have increased about twice as fast as ocean temperatures (0.25 °C per decade against 0.13 °C per decade). Ocean temperatures increase more slowly than land temperatures because of the larger effective heat capacity of the oceans and because the ocean loses more heat by evaporation. Since the beginning of industrialisation the temperature difference between the hemispheres has increased due to melting of sea ice and snow in the North. Average arctic temperatures have been increasing at almost twice the rate of the rest of the world in the past 100 years; however arctic temperatures are also highly variable. Although more greenhouse gases are emitted in the Northern than Southern Hemisphere this does not contribute to the difference in warming because the major greenhouse gases persist long enough to mix between hemispheres.

The thermal inertia of the oceans and slow responses of other indirect effects mean that climate can take centuries or longer to adjust to changes in forcing. Climate commitment studies indicate that even if greenhouse gases were stabilized at year 2000 levels, a further warming of about 0.5 °C (0.9 °F) would still occur.

Global temperature is subject to short-term fluctuations that overlay long-term trends and can temporarily mask them. The relative stability in surface temperature from 2002 to 2009, which has been dubbed the global warming hiatus by the media and some scientists, is consistent with such an episode. 2015 updates to account for differing methods of measuring ocean surface temperature measurements show a positive trend over the recent decade.

Warmest Years

15 of the top 16 warmest years have occurred since 2000. While record-breaking years can attract considerable public interest, individual years are less significant than the overall trend. So some climatologists have criticized the attention that the popular press gives to "warmest year" statistics; for example, Gavin Schmidt stated "the long-term trends or the expected sequence of records are far more important than whether any single year is a record or not."

2015 was not only the warmest year on record, it broke the record by the largest margin by which the record has been broken. 2015 was the 39th consecutive year with above-average temperatures. Ocean oscillations like El Niño Southern Oscillation (ENSO) can affect global average temperatures, for example, 1998 temperatures were significantly enhanced by strong El Niño conditions. 1998 remained the warmest year until 2005 and 2010 and the temperature of both of these years was enhanced by El Niño periods. The large margin by which 2015 is the warmest year is also attributed to another strong El Niño. However, 2014 was ENSO neutral. According to NOAA and NASA, 2015 had the warmest respective months on record for 10 out of the 12 months. The average temperature around the globe was 1.62°F (0.90°C) or 20% above the twentieth century average. In a first, December 2015 was also the first month to ever reach a temperature 2 degrees Fahrenheit above normal for the planet.

Initial Causes of Temperature Changes (External Forcings)

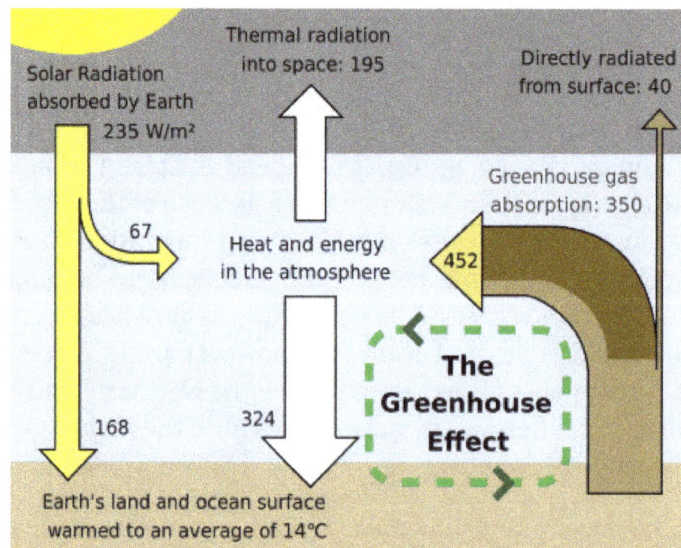

Greenhouse effect schematic showing energy flows between space, the atmosphere, and Earth's surface. Energy exchanges are expressed in watts per square meter (W/m²).

This graph, known as the Keeling Curve, documents the increase of atmospheric carbon dioxide concentrations from 1958–2015. Monthly CO_2 measurements display seasonal oscillations in an upward trend; each year's maximum occurs during the Northern Hemisphere's late spring, and declines during its growing season as plants remove some atmospheric CO_2.

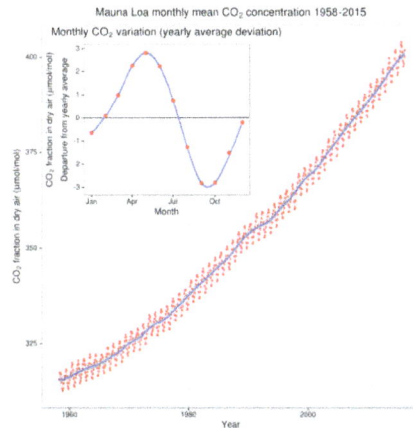

The climate system can warm or cool in response to changes in *external forcings*. These are "external" to the climate system but not necessarily external to Earth. Examples of external forcings include changes in atmospheric composition (e.g., increased concentrations of greenhouse gases), solar luminosity, volcanic eruptions, and variations in Earth's orbit around the Sun.

Greenhouse Gases

The greenhouse effect is the process by which absorption and emission of infrared radiation by gases in a planet's atmosphere warm its lower atmosphere and surface. It was proposed by Joseph Fourier in 1824, discovered in 1860 by John Tyndall, was first investigated quantitatively by Svante Arrhenius in 1896, and was developed in the 1930s through 1960s by Guy Stewart Callendar.

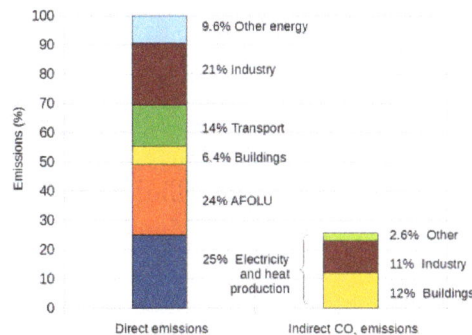

Annual world greenhouse gas emissions, in 2010, by sector.

Percentage share of global cumulative energy-related CO_2 emissions between 1751 and 2012 across different regions.

On Earth, naturally occurring amounts of greenhouse gases cause air temperature near the surface to be about 33 °C (59 °F) warmer than it would be in their absence.[d] Without the Earth's atmosphere, the Earth's average temperature would be well below the freezing temperature of water. The major greenhouse gases are water vapor, which causes about 36–70% of the greenhouse effect; carbon dioxide (CO_2), which causes 9–26%; methane (CH_4), which causes 4–9%; and ozone (O_3), which causes 3–7%. Clouds also affect the radiation balance through cloud forcings similar to greenhouse gases.

Human activity since the Industrial Revolution has increased the amount of greenhouse gases in the atmosphere, leading to increased radiative forcing from CO_2, methane, tropospheric ozone, CFCs and nitrous oxide. According to work published in 2007, the concentrations of CO_2 and methane have increased by 36% and 148% respectively since 1750. These levels are much higher than at any time during the last 800,000 years, the period for which reliable data has been extracted from ice cores. Less direct geological evidence indicates that CO_2 values higher than this were last seen about 20 million years ago.

Fossil fuel burning has produced about three-quarters of the increase in CO_2 from human activity over the past 20 years. The rest of this increase is caused mostly by changes in land-use, particularly deforestation. Another significant non-fuel source of anthropogenic CO_2 emissions is the calcination of limestone for clinker production, a chemical process which releases CO_2. Estimates of global CO_2 emissions in 2011 from fossil fuel combustion, including cement production and gas flaring, was 34.8 billion tonnes (9.5 ± 0.5 PgC), an increase of 54% above emissions in 1990. Coal burning was responsible for 43% of the total emissions, oil 34%, gas 18%, cement 4.9% and gas flaring 0.7%

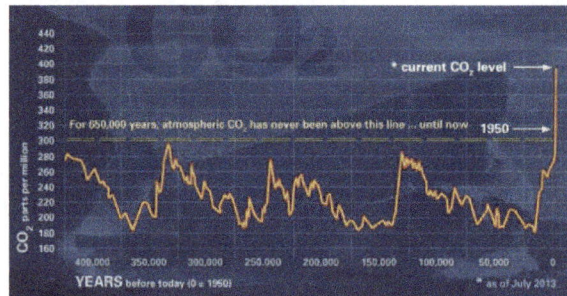

Atmospheric CO_2 concentration from 650,000 years ago to near present, using ice core proxy data and direct measurements.

In May 2013, it was reported that readings for CO_2 taken at the world's primary benchmark site in Mauna Loa surpassed 400 ppm. According to professor Brian Hoskins, this is likely the first time CO_2 levels have been this high for about 4.5 million years. Monthly global CO_2 concentrations exceeded 400 ppm in March 2015, probably for the first time in several million years. On 12 November 2015, NASA scientists reported that human-made carbon dioxide continues to increase above levels not seen in hundreds of thousands of years: currently, about half of the carbon dioxide released from the burning of fossil fuels is not absorbed by vegetation and the oceans and remains in the atmosphere.

Over the last three decades of the twentieth century, gross domestic product per capita and population growth were the main drivers of increases in greenhouse gas emissions. CO_2 emissions are continuing to rise due to the burning of fossil fuels and land-use change. Emissions can be attributed to different regions. Attributions of emissions due to land-use change are subject to considerable uncertainty.

Emissions scenarios, estimates of changes in future emission levels of greenhouse gases, have been projected that depend upon uncertain economic, sociological, technological, and natural developments. In most scenarios, emissions continue to rise over the century, while in a few, emissions are reduced. Fossil fuel reserves are abundant, and will not limit carbon emissions in the 21st century.

Emission scenarios, combined with modelling of the carbon cycle, have been used to produce estimates of how atmospheric concentrations of greenhouse gases might change in the future. Using the six IPCC SRES "marker" scenarios, models suggest that by the year 2100, the atmospheric concentration of CO_2 could range between 541 and 970 ppm. This is 90–250% above the concentration in the year 1750.

The popular media and the public often confuse global warming with ozone depletion, i.e., the destruction of stratospheric ozone (e.g., the ozone layer) by chlorofluorocarbons. Although there are a few areas of linkage, the relationship between the two is not strong. Reduced stratospheric ozone has had a slight cooling influence on surface temperatures, while increased tropospheric ozone has had a somewhat larger warming effect.

Aerosols and Soot

Ship tracks can be seen as lines in these clouds over the Atlantic Ocean on the east coast of the United States. Atmospheric particles from these and other sources could have a large effect on climate through the aerosol indirect effect.

Global dimming, a gradual reduction in the amount of global direct irradiance at the Earth's surface, was observed from 1961 until at least 1990. Solid and liquid particles known as *aerosols*, produced by volcanoes and human-made pollutants, are thought to be the main cause of this dimming. They exert a cooling effect by increasing the reflection of incoming sunlight. The effects of the products of fossil fuel combustion – CO_2 and aerosols – have partially offset one another in recent decades, so that net warming has been due to the increase in non-CO_2 greenhouse gases such as methane. Radiative forcing due to aerosols is temporally limited due to the processes that remove aerosols from the atmosphere. Removal by clouds and precipitation gives tropospheric aerosols an atmospheric lifetime of only about a week, while stratospheric aerosols can remain for a few years. Carbon dioxide has a lifetime of a century or more, and as such, changes in aerosols will only delay climate changes due to carbon dioxide. Black carbon is second only to carbon dioxide for its contribution to global warming.

In addition to their direct effect by scattering and absorbing solar radiation, aerosols have indirect effects on the Earth's radiation budget. Sulfate aerosols act as cloud condensation nuclei and thus lead to clouds that have more and smaller cloud droplets. These clouds reflect solar radiation more efficiently than clouds with fewer and larger droplets, a phenomenon known as the Twomey effect. This effect also causes droplets to be of more uniform size, which reduces growth of raindrops and makes the cloud more reflective to incoming sunlight, known as the Albrecht effect. Indirect effects are most noticeable in marine stratiform clouds, and have very little radiative effect on convective clouds. Indirect effects of aerosols represent the largest uncertainty in radiative forcing.

Soot may either cool or warm Earth's climate system, depending on whether it is airborne or deposited. Atmospheric soot directly absorbs solar radiation, which heats the atmosphere and cools the surface. In isolated areas with high soot production, such as rural India, as much as 50% of surface warming due to greenhouse gases may be masked by atmospheric brown clouds. When deposited, especially on glaciers or on ice in arctic regions, the lower surface albedo can also directly heat the surface. The influences of atmospheric particles, including black carbon, are most pronounced in the tropics and sub-tropics, particularly in Asia, while the effects of greenhouse gases are dominant in the extratropics and southern hemisphere.

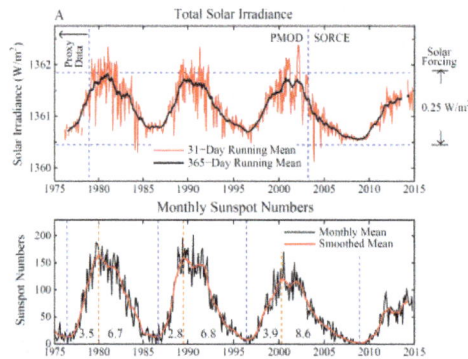

Changes in Total Solar Irradiance (TSI) and monthly sunspot numbers since the mid-1970s.

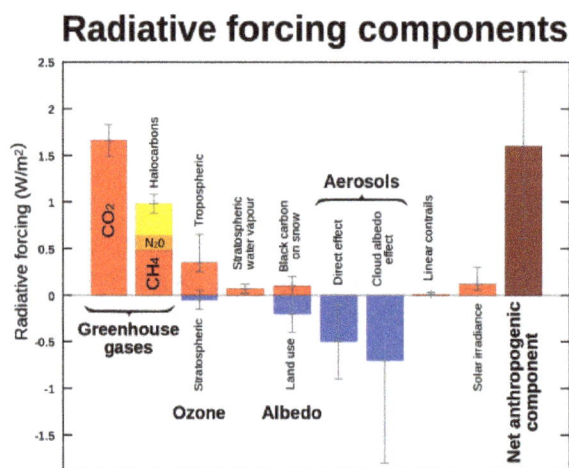

Contribution of natural factors and human activities to radiative forcing of climate change. Radiative forcing values are for the year 2005, relative to the pre-industrial era (1750). The contribution of solar irradiance to radiative forcing is 5% the value of the combined radiative forcing due to increases in the atmospheric concentrations of carbon dioxide, methane and nitrous oxide.

Solar Activity

Since 1978, solar irradiance has been measured by satellites. These measurements indicate that the Sun's radiative output has not increased since 1978, so the warming during the past 30 years cannot be attributed to an increase in solar energy reaching the Earth.

Climate models have been used to examine the role of the Sun in recent climate change. Models are unable to reproduce the rapid warming observed in recent decades when they only take into account variations in solar output and volcanic activity. Models are, however, able to simulate the observed 20th century changes in temperature when they include all of the most important external forcings, including human influences and natural forcings.

Another line of evidence against solar variations having caused recent climate change comes from looking at how temperatures at different levels in the Earth's atmosphere have changed. Models and observations show that greenhouse warming results in warming of the lower atmosphere (the troposphere) but cooling of the upper atmosphere (the stratosphere). Depletion of the ozone layer by chemical refrigerants has also resulted in a strong cooling effect in the stratosphere. If solar variations were responsible for observed warming, warming of both the troposphere and stratosphere would be expected.

Variations in Earth's Orbit

The tilt of the Earth's axis and the shape of its orbit around the Sun vary slowly over tens of thousands of years and are a natural source of climate change, by changing the seasonal and latitudinal distribution of solar insolation.

During the last few thousand years, this phenomenon contributed to a slow cooling trend at high latitudes of the Northern Hemisphere during summer, a trend that was reversed by greenhouse-gas-induced warming during the 20th century.

Variations in orbital cycles may initiate a new glacial period in the future, though the timing of this depends on greenhouse gas concentrations as well as the orbital forcing. A new glacial period is not expected within the next 50,000 years if atmospheric CO_2 concentration remains above 300 ppm.

Feedback

Sea ice, shown here in Nunavut, in northern Canada, reflects more sunshine, while open ocean absorbs more, accelerating melting.

The climate system includes a range of *feedbacks*, which alter the response of the system to changes in external forcings. Positive feedbacks increase the response of the climate system to an initial forcing, while negative feedbacks reduce it.

There are a range of feedbacks in the climate system, including water vapor, changes in ice-albedo (snow and ice cover affect how much the Earth's surface absorbs or reflects incoming sunlight), clouds, and changes in the Earth's carbon cycle (e.g., the release of carbon from soil). The main negative feedback is the energy the Earth's surface radiates into space as infrared radiation. According to the Stefan-Boltzmann law, if the absolute temperature (as measured in kelvin) doubles, radiated energy increases by a factor of 16 (2 to the 4th power).

Feedbacks are an important factor in determining the sensitivity of the climate system to increased atmospheric greenhouse gas concentrations. Other factors being equal, a higher *climate sensitivity* means that more warming will occur for a given increase in greenhouse gas forcing. Uncertainty over the effect of feedbacks is a major reason why different climate models project different magnitudes of warming for a given forcing scenario. More research is needed to understand the role of clouds and carbon cycle feedbacks in climate projections.

The IPCC projections previously mentioned span the "likely" range (greater than 66% probability, based on expert judgement) for the selected emissions scenarios. However, the IPCC's projections do not reflect the full range of uncertainty. The lower end of the "likely" range appears to be better constrained than the upper end.

Climate Models

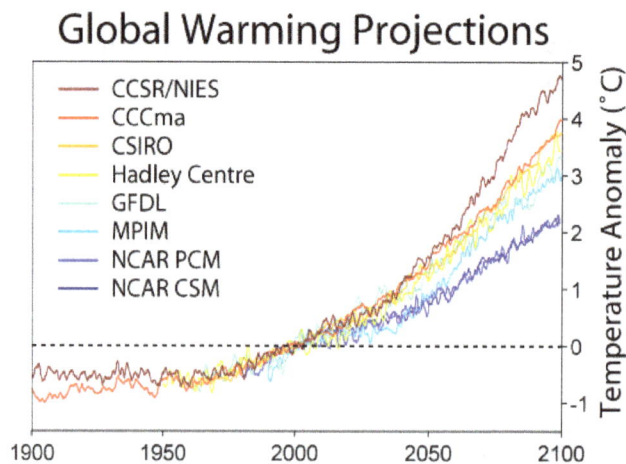

Calculations of global warming prepared in or before 2001 from a range of climate models under the SRES A2 emissions scenario, which assumes no action is taken to reduce emissions and regionally divided economic development.

Projected change in annual mean surface air temperature from the late 20th century to the middle 21st century, based on a medium emissions scenario (SRES A1B). This scenario assumes that no future policies are adopted to limit greenhouse gas emissions. Image credit: NOAA GFDL.

A climate model is a representation of the physical, chemical and biological processes that affect the climate system. Such models are based on scientific disciplines such as fluid dynamics and thermodynamics as well as physical processes such as radiative transfer. The models may be used to predict a range of variables such as local air movement, temperature, clouds, and other atmospheric properties; ocean temperature, salt content, and circulation; ice cover on land and sea; the transfer of heat and moisture from soil and vegetation to the atmosphere; and chemical and biological processes, among others.

Although researchers attempt to include as many processes as possible, simplifications of the actual climate system are inevitable because of the constraints of available computer power and limitations in knowledge of the climate system. Results from models can also vary due to different greenhouse gas inputs and the model's climate sensitivity. For example, the uncertainty in IPCC's 2007 projections is caused by (1) the use of multiple models with differing sensitivity to greenhouse gas concentrations, (2) the use of differing estimates of humanity's future greenhouse gas emissions, (3) any additional emissions from climate feedbacks that were not included in the models IPCC used to prepare its report, i.e., greenhouse gas releases from permafrost.

The models do not assume the climate will warm due to increasing levels of greenhouse gases. Instead the models predict how greenhouse gases will interact with radiative transfer and other physical processes. Warming or cooling is thus a result, not an assumption, of the models.

Clouds and their effects are especially difficult to predict. Improving the models' representation of clouds is therefore an important topic in current research. Another prominent research topic is expanding and improving representations of the carbon cycle.

Models are also used to help investigate the causes of recent climate change by comparing the observed changes to those that the models project from various natural and human causes. Although these models do not unambiguously attribute the warming that occurred from approximately 1910 to 1945 to either natural variation or human effects, they do indicate that the warming since 1970 is dominated by anthropogenic greenhouse gas emissions.

The physical realism of models is tested by examining their ability to simulate contemporary or past climates. Climate models produce a good match to observations of global temperature changes over the last century, but do not simulate all aspects of climate. Not all effects of global warming are accurately predicted by the climate models used by the IPCC. Observed Arctic shrinkage has been faster than that predicted. Precipitation increased proportionally to atmospheric humidity, and hence significantly faster than global climate models predict. Since 1990, sea level has also risen considerably faster than models predicted it would.

Observed and Expected Environmental Effects

Anthropogenic forcing has likely contributed to some of the observed changes, including sea level rise, changes in climate extremes (such as the number of warm and cold days), declines in Arctic sea ice extent, glacier retreat, and greening of the Sahara.

During the 21st century, glaciers and snow cover are projected to continue their widespread retreat. Projections of declines in Arctic sea ice vary. Recent projections suggest that Arctic summers could be ice-free (defined as ice extent less than 1 million square km) as early as 2025-2030.

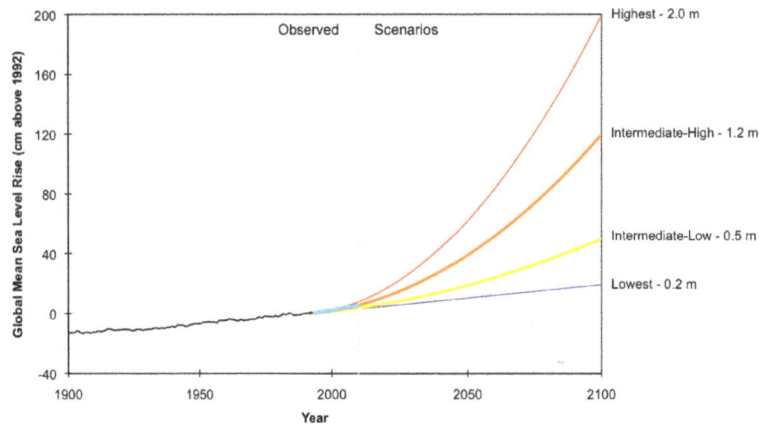

Projections of global mean sea level rise by Parris and others. Probabilities have not been assigned to these projections. Therefore, none of these projections should be interpreted as a "best estimate" of future sea level rise. Image credit: NOAA.

"Detection" is the process of demonstrating that climate has changed in some defined statistical sense, without providing a reason for that change. Detection does not imply attribution of the detected change to a particular cause. "Attribution" of causes of climate change is the process of establishing the most likely causes for the detected change with some defined level of confidence. Detection and attribution may also be applied to observed changes in physical, ecological and social systems.

Extreme Weather

Changes in regional climate are expected to include greater warming over land, with most warming at high northern latitudes, and least warming over the Southern Ocean and parts of the North Atlantic Ocean.

Future changes in precipitation are expected to follow existing trends, with reduced precipitation over subtropical land areas, and increased precipitation at subpolar latitudes and some equatorial regions. Projections suggest a probable increase in the frequency and severity of some extreme weather events, such as heat waves.

A 2015 study published in *Nature Climate Change*, states:

> "About 18% of the moderate daily precipitation extremes over land are attributable to the observed temperature increase since pre-industrial times, which in turn primarily results from human influence. For 2 °C of warming the fraction of precipitation extremes attributable to human influence rises to about 40%. Likewise, today about 75% of the moderate daily hot extremes over land are attributable to warming. It is the most rare and extreme events for which the largest fraction is anthropogenic, and that contribution increases non-linearly with further warming."

Data analysis of extreme events from 1960 till 2010 suggests that droughts and heat waves appear simultaneously with increased frequency. Extremely wet or dry events within the monsoon period have increased since 1980.

Sea Level Rise

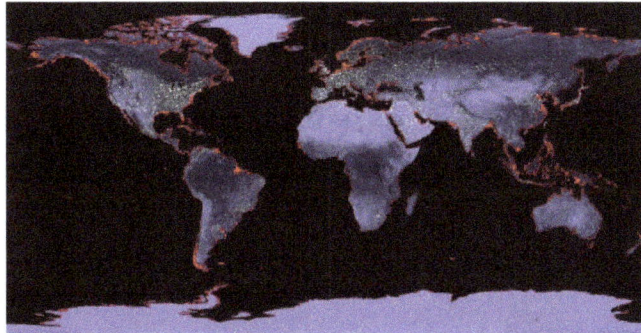

Map of the Earth with a six-meter sea level rise represented in red.

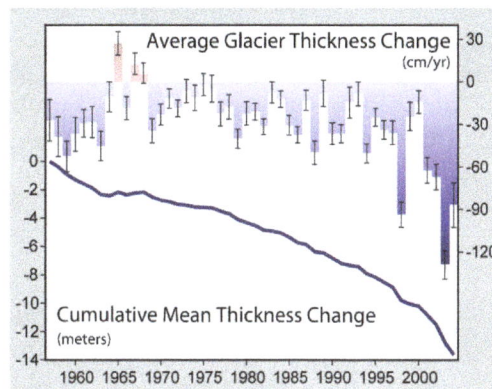

Sparse records indicate that glaciers have been retreating since the early 1800s. In the 1950s measurements began that allow the monitoring of glacial mass balance, reported to the World Glacier Monitoring Service (WGMS) and the National Snow and Ice Data Center (NSIDC).

The sea level rise since 1993 has been estimated to have been on average 2.6 mm and 2.9 mm per year ± 0.4 mm. Additionally, sea level rise has accelerated from 1995 to 2015. Over the 21st century, the IPCC projects for a high emissions scenario, that global mean sea level could rise by 52–98 cm. The IPCC's projections are conservative, and may underestimate future sea level rise. Other estimates suggest that for the same period, global mean sea level could rise by 0.2 to 2.0 m (0.7–6.6 ft), relative to mean sea level in 1992.

Widespread coastal flooding would be expected if several degrees of warming is sustained for millennia. For example, sustained global warming of more than 2 °C (relative to pre-industrial levels) could lead to eventual sea level rise of around 1 to 4 m due to thermal expansion of sea water and the melting of glaciers and small ice caps. Melting of the Greenland ice sheet could contribute an additional 4 to 7.5 m over many thousands of years. It has been estimated that we are already committed to a sea-level rise of approximately 2.3 meters for each degree of temperature rise within the next 2,000 years.

Warming beyond the 2 °C target would potentially lead to rates of sea-level rise dominated by ice loss from Antarctica. Continued CO_2 emissions from fossil sources could cause additional tens of meters of sea level rise, over the next millennia and eventually ultimately eliminate the entire Antarctic ice sheet, causing about 58 meters of sea level rise.

Ecological Systems

In terrestrial ecosystems, the earlier timing of spring events, as well as poleward and upward shifts in plant and animal ranges, have been linked with high confidence to recent warming. Future climate change is expected to affect particular ecosystems, including tundra, mangroves, and coral reefs. It is expected that most ecosystems will be affected by higher atmospheric CO_2 levels, combined with higher global temperatures. Overall, it is expected that climate change will result in the extinction of many species and reduced diversity of ecosystems.

Increases in atmospheric CO_2 concentrations have led to an increase in ocean acidity. Dissolved CO_2 increases ocean acidity, measured by lower pH values. Between 1750 and 2000, surface-ocean pH has decreased by ≈ 0.1, from ≈ 8.2 to ≈ 8.1. Surface-ocean pH has probably not been below ≈ 8.1 during the past 2 million years. Projections suggest that surface-ocean pH could decrease by an additional 0.3–0.4 units by 2100. Future ocean acidification could threaten coral reefs, fisheries, protected species, and other natural resources of value to society.

Ocean deoxygenation is projected to increase hypoxia by 10%, and triple suboxic waters (oxygen concentrations 98% less than the mean surface concentrations), for each 1 °C of upper ocean warming.

Long-term Effects

On the timescale of centuries to millennia, the magnitude of global warming will be determined primarily by anthropogenic CO_2 emissions. This is due to carbon dioxide's very long lifetime in the atmosphere.

Stabilizing the global average temperature would require large reductions in CO_2 emissions, as well as reductions in emissions of other greenhouse gases such as methane and nitrous oxide. Emissions of CO_2 would need to be reduced by more than 80% relative to their peak level. Even if this were achieved, global average temperatures would remain close to their highest level for many centuries.

Long-term effects also include a response from the Earth's crust, due to ice melting and deglaciation, in a process called post-glacial rebound, when land masses are no longer depressed by the weight of ice. This could lead to landslides and increased seismic and volcanic activities. Tsunamis could be generated by submarine landslides caused by warmer ocean water thawing ocean-floor permafrost or releasing gas hydrates. Some world regions, such as the French Alps, already show signs of an increase in landslide frequency.

Large-scale and Abrupt Impacts

Climate change could result in global, large-scale changes in natural and social systems. Examples include the possibility for the Atlantic Meridional Overturning Circulation to slow- or shutdown, which in the instance of a shutdown would change weather in Europe and North America considerably, ocean acidification caused by increased atmospheric concentrations of carbon dioxide, and the long-term melting of ice sheets, which contributes to sea level rise.

Some large-scale changes could occur abruptly, i.e., over a short time period, and might also be irreversible. Examples of abrupt climate change are the rapid release of methane and carbon dioxide from permafrost, which would lead to amplified global warming, or the shutdown of thermohaline circulation. Scientific understanding of abrupt climate change is generally poor. The probability of abrupt change for some climate related feedbacks may be low. Factors that may increase the probability of abrupt climate change include higher magnitudes of global warming, warming that occurs more rapidly, and warming that is sustained over longer time periods.

Observed and Expected Effects on Social Systems

The effects of climate change on human systems, mostly due to warming or shifts in precipitation patterns, or both, have been detected worldwide. Production of wheat and maize globally has been impacted by climate change. While crop production has increased in some mid-latitude regions such as the UK and Northeast China, economic losses due to extreme weather events have increased globally. There has been a shift from cold- to heat-related mortality in some regions as a result of warming. Livelihoods of indigenous peoples of the Arctic have been altered by climate change, and there is emerging evidence of climate change impacts on livelihoods of indigenous peoples in other regions. Regional impacts of climate change are now observable at more locations than before, on all continents and across ocean regions.

The future social impacts of climate change will be uneven. Many risks are expected to increase with higher magnitudes of global warming. All regions are at risk of experiencing negative impacts. Low-latitude, less developed areas face the greatest risk. A study from 2015 concluded that economic growth (gross domestic product) of poorer countries is much more impaired with projected future climate warming, than previously thought.

A meta-analysis of 56 studies concluded in 2014 that each degree of temperature rise will increase violence by up to 20%, which includes fist fights, violent crimes, civil unrest or wars.

Examples of impacts include:

- *Food*: Crop production will probably be negatively affected in low latitude countries, while effects at northern latitudes may be positive or negative. Global warming of around 4.6 °C relative to pre-industrial levels could pose a large risk to global and regional food security.

- *Health*: Generally impacts will be more negative than positive. Impacts include: the effects of extreme weather, leading to injury and loss of life; and indirect effects, such as undernutrition brought on by crop failures.

Habitat Inundation

In small islands and mega deltas, inundation as a result of sea level rise is expected to threaten vital infrastructure and human settlements. This could lead to issues of homelessness in countries with low-lying areas such as Bangladesh, as well as statelessness for populations in countries such as the Maldives and Tuvalu.

Economy

Estimates based on the IPCC A1B emission scenario from additional CO_2 and CH_4 greenhouse gases released from permafrost, estimate associated impact damages by US$43 trillion.

Infrastructure

Continued permafrost degradation will likely result in unstable infrastructure in Arctic regions, or Alaska before 2100. Thus, impacting roads, pipelines and buildings, as well as water distribution, and cause slope failures.

Possible Responses to Global Warming

Mitigation

The graph on the right shows three "pathways" to meet the UNFCCC's 2 °C target, labelled "global technology", "decentralised solutions", and "consumption change". Each pathway shows how various measures (e.g., improved energy efficiency, increased use of renewable energy) could contribute to emissions reductions. Image credit: PBL Netherlands Environmental Assessment Agency.

Mitigation of climate change are actions to reduce greenhouse gas emissions, or enhance the capacity of carbon sinks to absorb GHGs from the atmosphere. There is a large potential for future reductions in emissions by a combination of activities, including: energy conservation and increased energy efficiency; the use of low-carbon energy technologies, such as renewable energy, nuclear energy, and carbon capture and storage; and enhancing carbon sinks through, for example, reforestation and preventing deforestation. A 2015 report by Citibank concluded that transitioning to a low carbon economy would yield positive return on investments.

Near- and long-term trends in the global energy system are inconsistent with limiting global warming at below 1.5 or 2 °C, relative to pre-industrial levels. Pledges made as part of the Cancún agreements are broadly consistent with having a likely chance (66 to 100% probability) of limiting global warming (in the 21st century) at below 3 °C, relative to pre-industrial levels.

In limiting warming at below 2 °C, more stringent emission reductions in the near-term would allow for less rapid reductions after 2030. Many integrated models are unable to meet the 2 °C target if pessimistic assumptions are made about the availability of mitigation technologies.

Adaptation

Other policy responses include adaptation to climate change. Adaptation to climate change may be planned, either in reaction to or anticipation of climate change, or spontaneous, i.e., without government intervention. Planned adaptation is already occurring on a limited basis. The barriers, limits, and costs of future adaptation are not fully understood.

A concept related to adaptation is *adaptive capacity*, which is the ability of a system (human, natural or managed) to adjust to climate change (including climate variability and extremes) to moderate potential damages, to take advantage of opportunities, or to cope with consequences. Unmitigated climate change (i.e., future climate change without efforts to limit greenhouse gas emissions) would, in the long term, be likely to exceed the capacity of natural, managed and human systems to adapt.

Environmental organizations and public figures have emphasized changes in the climate and the risks they entail, while promoting adaptation to changes in infrastructural needs and emissions reductions.

Climate Engineering

Climate engineering (sometimes called *geoengineering* or *climate intervention*) is the deliberate modification of the climate. It has been investigated as a possible response to global warming, e.g. by NASA and the Royal Society. Techniques under research fall generally into the categories solar radiation management and carbon dioxide removal, although various other schemes have been suggested. A study from 2014 investigated the most common climate engineering methods and concluded they are either ineffective or have potentially severe side effects and cannot be stopped without causing rapid climate change.

Discourse About Global Warming

Political Discussion

Most countries in the world are parties to the United Nations Framework Convention on Climate Change (UNFCCC). The ultimate objective of the Convention is to prevent dangerous human interference of the climate system. As stated in the Convention, this requires that GHG concentrations are stabilized in the atmosphere at a level where ecosystems can adapt naturally to climate change, food production is not threatened, and economic development can proceed in a sustainable fashion. The Framework Convention was agreed in 1992, but since then, global emissions have risen.

During negotiations, the G77 (a lobbying group in the United Nations representing 133 developing nations) pushed for a mandate requiring developed countries to "[take] the lead" in reducing their emissions. This was justified on the basis that: the developed world's emissions had contributed most to the cumulation of GHGs in the atmosphere; per-capita emissions (i.e., emissions per head of population) were still relatively low in developing countries; and the emissions of developing countries would grow to meet their development needs.

This mandate was sustained in the Kyoto Protocol to the Framework Convention, which entered into legal effect in 2005. In ratifying the Kyoto Protocol, most developed countries accepted legally binding commitments to limit their emissions. These first-round commitments expired in 2012. United States President George W. Bush rejected the treaty on the basis that "it exempts 80% of the world, including major population centers such as China and India, from compliance, and would cause serious harm to the US economy."

At the 15th UNFCCC Conference of the Parties, held in 2009 at Copenhagen, several UNFCCC Parties produced the Copenhagen Accord. Parties associated with the Accord (140 countries, as of November 2010) aim to limit the future increase in global mean temperature to below 2 °C. The 16th Conference of the Parties (COP16) was held at Cancún in 2010. It produced an agreement, not a binding treaty, that the Parties should take urgent action to reduce greenhouse gas emissions to meet a goal of limiting global warming to 2 °C above pre-industrial temperatures. It also recognized the need to consider strengthening the goal to a global average rise of 1.5 °C.

Scientific Discussion

There is continuing discussion through published peer-reviewed scientific papers, which are assessed by scientists working in the relevant fields taking part in the Intergovernmental Panel on

Climate Change. The scientific consensus as of 2013 stated in the IPCC Fifth Assessment Report is that it "is extremely likely that human influence has been the dominant cause of the observed warming since the mid-20th century". A 2008 report by the U.S. National Academy of Sciences stated that most scientists by then agreed that observed warming in recent decades was primarily caused by human activities increasing the amount of greenhouse gases in the atmosphere. In 2005 the Royal Society stated that while the overwhelming majority of scientists were in agreement on the main points, some individuals and organisations opposed to the consensus on urgent action needed to reduce greenhouse gas emissions have tried to undermine the science and work of the IPCC. National science academies have called on world leaders for policies to cut global emissions.

In the scientific literature, there is a strong consensus that global surface temperatures have increased in recent decades and that the trend is caused mainly by human-induced emissions of greenhouse gases. No scientific body of national or international standing disagrees with this view.

Discussion by the Public and in Popular Media

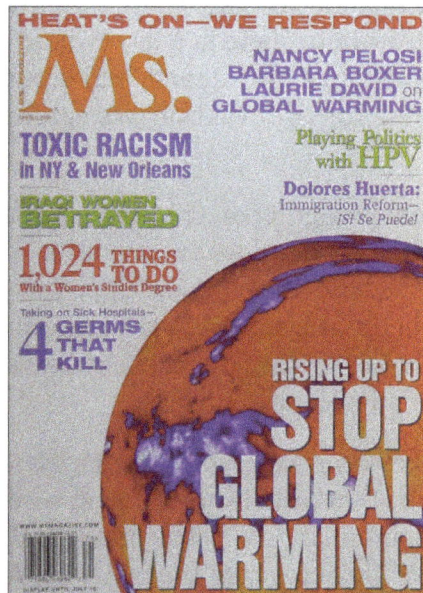

Global warming was the cover story in this 2007 issue of *Ms. magazine*

The global warming controversy refers to a variety of disputes, substantially more pronounced in the popular media than in the scientific literature, regarding the nature, causes, and consequences of global warming. The disputed issues include the causes of increased global average air temperature, especially since the mid-20th century, whether this warming trend is unprecedented or within normal climatic variations, whether humankind has contributed significantly to it, and whether the increase is completely or partially an artifact of poor measurements. Additional disputes concern estimates of climate sensitivity, predictions of additional warming, and what the consequences of global warming will be.

From 1990 to 1997, right-wing conservative think tanks in the United States mobilized to challenge the legitimacy of global warming as a social problem. They challenged the scientific evidence, argued that global warming will have benefits, and asserted that proposed solutions would do more harm than good. Some people dispute aspects of climate change science. Organizations such as the libertarian Competitive Enterprise Institute, conservative commentators, and some companies such as ExxonMobil have challenged IPCC climate change scenarios, funded scientists who disagree with the scientific consensus, and provided their own projections of the economic cost of stricter controls. On the other hand, some fossil fuel companies have scaled back their efforts in recent years, or even called for policies to reduce global warming. Global oil companies have begun to acknowledge climate change exists and is caused by human activities and the burning of fossil fuels.

Surveys of Public Opinion

The world public, or at least people in economically advanced regions, became broadly aware of the global warming problem in the late 1980s. Polling groups began to track opinions on the subject, at first mainly in the United States. The longest consistent polling, by Gallup in the US, found relatively small deviations of 10% or so from 1998 to 2015 in opinion on the seriousness of global warming, but with increasing polarization between those concerned and those unconcerned.

The first major worldwide poll, conducted by Gallup in 2008-2009 in 127 countries, found that some 62% of people worldwide said they knew about global warming. In the advanced countries of North America, Europe and Japan, 90% or more knew about it (97% in the U.S., 99% in Japan); in less developed countries, especially in Africa, fewer than a quarter knew about it, although many had noticed local weather changes. Among those who knew about global warming, there was a wide variation between nations in belief that the warming was a result of human activities.

By 2010, with 111 countries surveyed, Gallup determined that there was a substantial decrease since 2007–08 in the number of Americans and Europeans who viewed global warming as a serious threat. In the US, just a little over half the population (53%) now viewed it as a serious concern for either themselves or their families; this was 10 points below the 2008 poll (63%). Latin America had the biggest rise in concern: 73% said global warming is a serious threat to their families. This global poll also found that people are more likely to attribute global warming to human activities than to natural causes, except in the US where nearly half (47%) of the population attributed global warming to natural causes.

A March–May 2013 survey by Pew Research Center for the People & the Press polled 39 countries about global threats. According to 54% of those questioned, global warming featured top of the perceived global threats. In a January 2013 survey, Pew found that 69% of Americans say there is solid evidence that the Earth's average temperature has got warmer over the past few decades, up six points since November 2011 and 12 points since 2009.

A 2010 survey of 14 industrialized countries found that skepticism about the danger of global warming was highest in Australia, Norway, New Zealand and the United States, in that order, correlating positively with per capita emissions of carbon dioxide.

Etymology

In the 1950s, research suggested increasing temperatures, and a 1952 newspaper reported "climate change". This phrase next appeared in a November 1957 report in *The Hammond Times* which described Roger Revelle's research into the effects of increasing human-caused CO_2 emissions on the greenhouse effect, "a large scale global warming, with radical climate changes may result". Both phrases were only used occasionally until 1975, when Wallace Smith Broecker published a scientific paper on the topic; "Climatic Change: Are We on the Brink of a Pronounced Global Warming?" The phrase began to come into common use, and in 1976 Mikhail Budyko's statement that "a global warming up has started" was widely reported. Other studies, such as a 1971 MIT report, referred to the human impact as "inadvertent climate modification", but an influential 1979 National Academy of Sciences study headed by Jule Charney followed Broecker in using *global warming* for rising surface temperatures, while describing the wider effects of increased CO_2 as *climate change.*

In 1986 and November 1987, NASA climate scientist James Hansen gave testimony to Congress on global warming. There were increasing heatwaves and drought problems in the summer of 1988, and when Hansen testified in the Senate on 23 June he sparked worldwide interest. He said: "global warming has reached a level such that we can ascribe with a high degree of confidence a cause and effect relationship between the greenhouse effect and the observed warming." Public attention increased over the summer, and *global warming* became the dominant popular term, commonly used both by the press and in public discourse.

In a 2008 NASA article on usage, Erik M. Conway defined *Global warming* as "the increase in Earth's average surface temperature due to rising levels of greenhouse gases", while *Climate change* was "a long-term change in the Earth's climate, or of a region on Earth." As effects such as changing patterns of rainfall and rising sea levels would probably have more impact than temperatures alone, he considered *global climate change* a more scientifically accurate term, and like the Intergovernmental Panel on Climate Change, the NASA website would emphasise this wider context.

Ozone Depletion

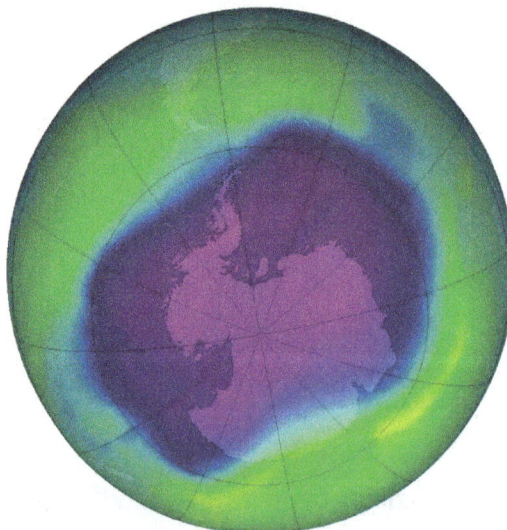

Image of the largest Antarctic ozone hole ever recorded (September 2006), over the Southern pole

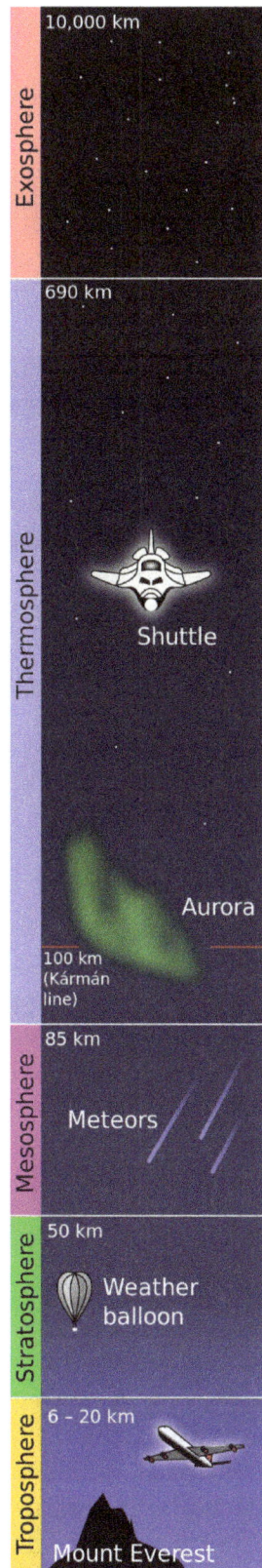

10,000 km

Exosphere

690 km

Thermosphere

Shuttle

Aurora

100 km
(Kármán
line)

85 km

Mesosphere

Meteors

50 km

Stratosphere

Weather
balloon

6 – 20 km

Troposphere

Mount Everest

Layers of the atmosphere (not to scale). The Earth's ozone layer is mainly found in the lower portion of the stratosphere from approximately 20 to 30 km (12 to 19 mi).

Ozone depletion describes two distinct but related phenomena observed since the late 1970s: a steady decline of about four percent in the total amount of ozone in Earth's stratosphere (the ozone layer), and a much larger springtime decrease in stratospheric ozone around Earth's polar regions. The latter phenomenon is referred to as the ozone hole. In addition to these well-known stratospheric phenomena, there are also springtime polar tropospheric ozone depletion events.

The details of polar ozone hole formation differ from that of mid-latitude thinning but the most important process in both is catalytic destruction of ozone by atomic halogens. The main source of these halogen atoms in the stratosphere is photodissociation of man-made halocarbon refrigerants, solvents, propellants, and foam-blowing agents (chlorofluorocarbon (CFCs), HCFCs, freons, halons). These compounds are transported into the stratosphere by winds after being emitted at the surface. Both types of ozone depletion were observed to increase as emissions of halocarbons increased.

CFCs and other contributory substances are referred to as ozone-depleting substances (ODS). Since the ozone layer prevents most harmful UVB wavelengths (280–315 nm) of ultraviolet light (UV light) from passing through the Earth's atmosphere, observed and projected decreases in ozone generated worldwide concern, leading to adoption of the Montreal Protocol that bans the production of CFCs, halons, and other ozone-depleting chemicals such as carbon tetrachloride and trichloroethane. It is suspected that a variety of biological consequences such as increases in sunburn, skin cancer, cataracts, damage to plants, and reduction of plankton populations in the ocean's photic zone may result from the increased UV exposure due to ozone depletion.

Ozone Cycle Overview

Three forms (or allotropes) of oxygen are involved in the ozone-oxygen cycle: oxygen atoms (O or atomic oxygen), oxygen gas (O_2 or diatomic oxygen), and ozone gas (O_3 or triatomic oxygen). Ozone is formed in the stratosphere when oxygen molecules photodissociate after intaking an ultraviolet photon whose wavelength is shorter than 240 nm. This converts a single O_2 into two atomic oxygen radicals. The atomic oxygen radicals then combine with separate O_2 molecules to create two O_3 molecules. These ozone molecules absorb ultraviolet (UV) light between 310 and 200 nm, following which ozone splits into a molecule of O_2 and an oxygen atom. The oxygen atom then joins up with an oxygen molecule to regenerate ozone. This is a continuing process that terminates when an oxygen atom "recombines" with an ozone molecule to make two O_2 molecules.

$$2 \, O_3 \rightarrow 3 \, O_2$$

The overall amount of ozone in the stratosphere is determined by a balance between photochemical production and recombination.

Ozone can be destroyed by a number of free radical catalysts, the most important of which are the hydroxyl radical (OH·), nitric oxide radical (NO·), chlorine radical (Cl·) and bromine radical (Br·). The dot is a common notation to indicate that all of these species have an unpaired electron and are thus extremely reactive. All of these have both natural and man-made sources; at the present time, most of the OH· and NO· in the stratosphere is of natural origin, but human activity has dramatically increased the levels of chlorine and bromine. These elements are found in certain stable organic compounds, especially chlorofluorocarbons, which may find their way to the stratosphere without being destroyed in the troposphere due to their low reactivity. Once in the stratosphere, the Cl and Br atoms are liberated from the parent compounds by the action of ultraviolet light, e.g.

CFCl3 + electromagnetic radiation → Cl· + ·CFCl2

The Cl and Br atoms can then destroy ozone molecules through a variety of catalytic cycles. In the simplest example of such a cycle, a chlorine atom reacts with an ozone molecule, taking an oxygen atom with it (forming ClO) and leaving a normal oxygen molecule. The chlorine monoxide (i.e., the ClO) can react with a second molecule of ozone (i.e., O3) to yield another chlorine atom and two molecules of oxygen. The chemical shorthand for these gas-phase reactions is:

- Cl· + O3 → ClO + O2: The chlorine atom changes an ozone molecule to ordinary oxygen

- ClO + O3 → Cl· + 2 O2: The ClO from the previous reaction destroys a second ozone molecule and recreates the original chlorine atom, which can repeat the first reaction and continue to destroy ozone.

The overall effect is a decrease in the amount of ozone, though the rate of these processes can be decreased by the effects of null cycles. More complicated mechanisms have been discovered that lead to ozone destruction in the lower stratosphere as well.

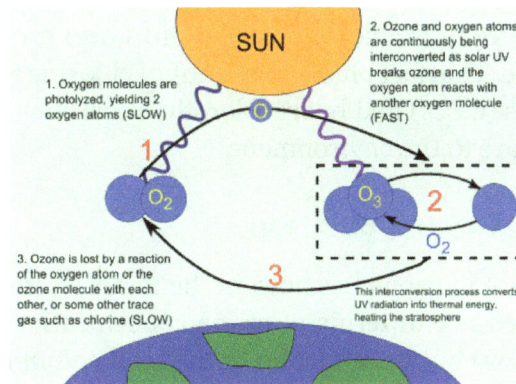

The ozone cycle

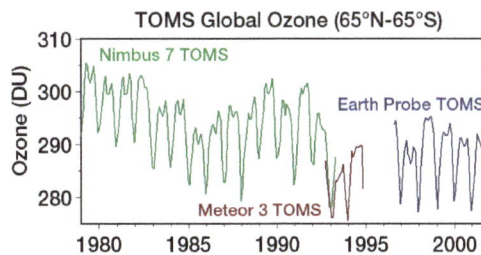

Global monthly average total ozone amount

A single chlorine atom would keep on destroying ozone (thus a catalyst) for up to two years (the time scale for transport back down to the troposphere) were it not for reactions that remove them from this cycle by forming reservoir species such as hydrogen chloride (HCl) and chlorine nitrate (ClONO2). On a per atom basis, bromine is even more efficient than chlorine at destroying ozone, but there is much less bromine in the atmosphere at present. As a result, both chlorine and bromine contribute significantly to overall ozone depletion. Laboratory studies have shown that fluorine and iodine atoms participate in analogous catalytic cycles. However, in the Earth's stratosphere, fluorine atoms react rapidly with water and methane to form strongly bound HF, while organic molecules containing iodine react so rapidly in the lower atmosphere that they do not reach the stratosphere in significant quantities.

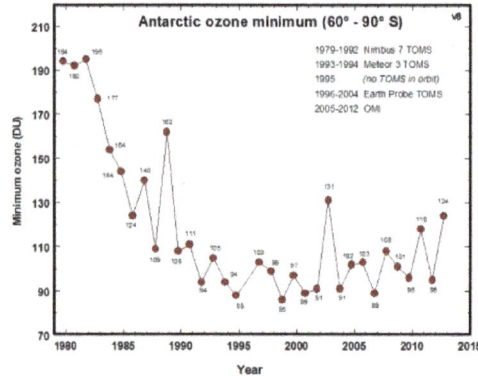

Lowest value of ozone measured by TOMS each year in the ozone hole

On average, a single chlorine atom is able to react with 100,000 ozone molecules before it is removed from the catalytic cycle. This fact plus the amount of chlorine released into the atmosphere yearly by chlorofluorocarbons (CFCs) and hydrochlorofluorocarbons (HCFCs) demonstrates how dangerous CFCs and HCFCs are to the environment.

Observations on Ozone Layer Depletion

The most-pronounced decrease in ozone has been in the lower stratosphere. However, the ozone hole is most usually measured not in terms of ozone concentrations at these levels (which are typically a few parts per million) but by reduction in the total *column ozone* above a point on the Earth's surface, which is normally expressed in Dobson units, abbreviated as "DU". Marked decreases in column ozone in the Antarctic spring and early summer compared to the early 1970s and before have been observed using instruments such as the Total Ozone Mapping Spectrometer (TOMS).

Reductions of up to 70 percent in the ozone column observed in the austral (southern hemispheric) spring over Antarctica and first reported in 1985 (Farman et al.) are continuing. Since the 1990s, Antarctic total column ozone in September and October continued to be 40–50 percent lower than pre-ozone-hole values. A gradual trend toward "healing" was reported in 2016.

In the Arctic, the amount lost is more variable year-to-year than in the Antarctic. The greatest Arctic declines, up to 30 percent, are in the winter and spring, when the stratosphere is coldest.

Reactions that take place on polar stratospheric clouds (PSCs) play an important role in enhancing ozone depletion. PSCs form more readily in the extreme cold of the Arctic and Antarctic stratosphere. This is why ozone holes first formed, and are deeper, over Antarctica. Early models failed to take PSCs into account and predicted a gradual global depletion, which is why the sudden Antarctic ozone hole was such a surprise to many scientists.

In middle latitudes, it is more accurate to speak of ozone depletion rather than holes. Total column ozone declined to about six percent below pre-1980 values between 1980 and 1996 for the mid-latitudes of 35–60°N and 35–60°S. In the northern mid-latitudes, it then increased from the minimum value by about two percent from 1996 to 2009 as regulations took effect and the amount of chlorine in the stratosphere decreased. In the Southern Hemisphere's mid-latitudes, total ozone remained constant over that time period. In the tropics, there are no significant trends, largely because halogen-containing compounds have not yet had time to break down and release chlorine and bromine atoms at tropical latitudes.

Large volcanic eruptions have been shown to have substantial albeit uneven ozone-depleting effects, as observed (for example) with the 1991 eruption of Mt. Pinotubo in the Philippines.

Ozone depletion also explains much of the observed reduction in stratospheric and upper tropospheric temperatures. The source of the warmth of the stratosphere is the absorption of UV radiation by ozone, hence reduced ozone leads to cooling. Some stratospheric cooling is also predicted from increases in greenhouse gases such as CO_2 and CFCs themselves; however the ozone-induced cooling appears to be dominant.

Predictions of ozone levels remain difficult, but the precision of models' predictions of observed values and the agreement among different modeling techniques have increased steadily. The World Meteorological Organization Global Ozone Research and Monitoring Project—Report No. 44 comes out strongly in favor of the Montreal Protocol, but notes that a UNEP 1994 Assessment overestimated ozone loss for the 1994–1997 period.

Chemicals in the Atmosphere

Cfcs And Related Compounds in the Atmosphere

Chlorofluorocarbons (CFCs) and other halogenated ozone depleting substances (ODS) are mainly responsible for man-made chemical ozone depletion. The total amount of effective halogens (chlorine and bromine) in the stratosphere can be calculated and are known as the equivalent effective stratospheric chlorine (EESC).

CFCs were invented by Thomas Midgley, Jr. in the 1920s. They were used in air conditioning and cooling units, as aerosol spray propellants prior to the 1970s, and in the cleaning processes of delicate electronic equipment. They also occur as by-products of some chemical processes. No significant natural sources have ever been identified for these compounds—their presence in the atmosphere is due almost entirely to human manufacture. As mentioned above, when such ozone-depleting chemicals reach the stratosphere, they are dissociated by ultraviolet light to release chlorine atoms. The chlorine atoms act as a catalyst, and each can break down tens of thousands of ozone molecules before being removed from the stratosphere. Given the longevity of CFC molecules, recovery times are measured in decades. It is calculated that a CFC molecule takes an average of about five to seven years to go from the ground level up to the upper atmosphere, and it can stay there for about a century, destroying up to one hundred thousand ozone molecules during that time.

1,1,1-Trichloro-2,2,2-trifluoroethane, also known as CFC-113a, is one of four man-made chemicals newly discovered in the atmosphere by a team at the University of East Anglia. CFC-113a is the

3

only known CFC whose abundance in the atmosphere is still growing. Its source remains a mystery, but illegal manufacturing is suspected by some. CFC-113a seems to have been accumulating unabated since 1960. Between 2010 and 2012, emissions of the gas jumped by 45 percent.

Computer Modeling

Scientists have been increasingly able to attribute the observed ozone depletion to the increase of man-made (anthropogenic) halogen compounds from CFCs by the use of complex chemistry transport models and their validation against observational data (e.g. SLIMCAT, CLaMS—Chemical Lagrangian Model of the Stratosphere). These models work by combining satellite measurements of chemical concentrations and meteorological fields with chemical reaction rate constants obtained in lab experiments. They are able to identify not only the key chemical reactions but also the transport processes that bring CFC photolysis products into contact with ozone.

Ozone Hole and its Causes

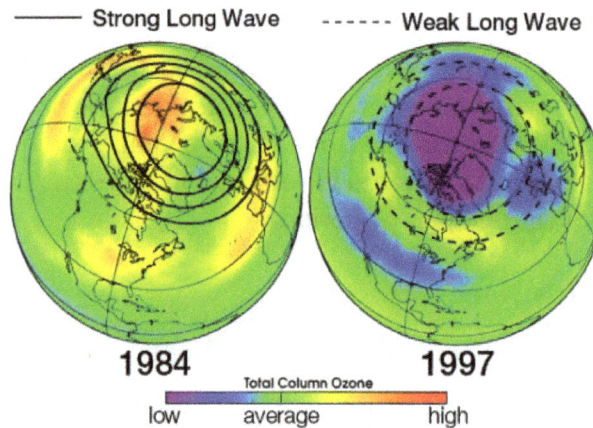

Ozone hole in North America during 1984 (abnormally warm reducing ozone depletion) and 1997 (abnormally cold resulting in increased seasonal depletion). Source: NASA

The Antarctic ozone hole is an area of the Antarctic stratosphere in which the recent ozone levels have dropped to as low as 33 percent of their pre-1975 values. The ozone hole occurs during the Antarctic spring, from September to early December, as strong westerly winds start to circulate around the continent and create an atmospheric container. Within this polar vortex, over 50 percent of the lower stratospheric ozone is destroyed during the Antarctic spring.

As explained above, the primary cause of ozone depletion is the presence of chlorine-containing source gases (primarily CFCs and related halocarbons). In the presence of UV light, these gases dissociate, releasing chlorine atoms, which then go on to catalyze ozone destruction. The Cl-catalyzed ozone depletion can take place in the gas phase, but it is dramatically enhanced in the presence of polar stratospheric clouds (PSCs).

These polar stratospheric clouds (PSC) form during winter, in the extreme cold. Polar winters are dark, consisting of 3 months without solar radiation (sunlight). The lack of sunlight contributes to a decrease in temperature and the polar vortex traps and chills air. Temperatures hover around or below −80 °C. These low temperatures form cloud particles. There are three types of PSC clouds—nitric acid trihydrate clouds, slowly cooling water-ice clouds, and rapid cooling water-ice (nacerous) clouds—provide surfaces for chemical reactions whose products will, in the spring lead to ozone destruction.

The photochemical processes involved are complex but well understood. The key observation is that, ordinarily, most of the chlorine in the stratosphere resides in "reservoir" compounds, primarily chlorine nitrate ($ClONO_2$) as well as stable end products such as HCl. The formation of end products essentially remove Cl from the ozone depletion process. The former sequester Cl, which can be later made available via absorption of light at shorter wavelengths than 400 nm. During the Antarctic winter and spring, however, reactions on the surface of the polar stratospheric cloud particles convert these "reservoir" compounds into reactive free radicals (Cl and ClO). The process by which the clouds remove NO_2 from the stratosphere by converting it to nitric acid in the PSC particles, which then are lost by sedimentation is called denitrification. This prevents newly formed ClO from being converted back into $ClONO_2$.

The role of sunlight in ozone depletion is the reason why the Antarctic ozone depletion is greatest during spring. During winter, even though PSCs are at their most abundant, there is no light over the pole to drive chemical reactions. During the spring, however, the sun comes out, providing energy to drive photochemical reactions and melt the polar stratospheric clouds, releasing considerable ClO, which drives the hole mechanism. Further warming temperatures near the end of spring break up the vortex around mid-December. As warm, ozone and NO_2-rich air flows in from lower latitudes, the PSCs are destroyed, the enhanced ozone depletion process shuts down, and the ozone hole closes.

Most of the ozone that is destroyed is in the lower stratosphere, in contrast to the much smaller ozone depletion through homogeneous gas phase reactions, which occurs primarily in the upper stratosphere.

Interest in Ozone Layer Depletion

Public misconceptions and misunderstandings of complex issues like the ozone depletion are common. The limited scientific knowledge of the public led to a confusion with global warming or the perception of global warming as a subset of the "ozone hole". In the beginning, classical green NGOs refrained from using CFC depletion for campaigning, as they assumed the topic was too complicated. They became active much later, e.g. in Greenpeace's support for a CFC-free fridge produced by the former East German company VEB dkk Scharfenstein.

The metaphors used in the CFC discussion (ozone shield, ozone hole) are not "exact" in the scientific sense. The "ozone hole" is more of a *depression*, less "a hole in the windshield". The ozone does not disappear through the layer, nor is there a uniform "thinning" of the ozone layer. However they resonated better with non-scientists and their concerns. The ozone hole was seen as a "hot issue" and imminent risk as lay people feared severe personal consequences such skin cancer, cataracts, damage to plants, and reduction of plankton populations in the ocean's photic zone. Not only on the policy level, ozone regulation compared to climate change fared much better in public opinion. Americans voluntarily switched away from aerosol sprays before legislation was enforced, while climate change failed to achieve comparable concern and public action. The sudden recognition in 1985 that there was a substantial "hole" was widely reported in the press. The especially rapid ozone depletion in Antarctica had previously been dismissed as a measurement error. Scientific consensus was established after regulation.

While the Antarctic ozone hole has a relatively small effect on global ozone, the hole has generated a great deal of public interest because:

- Many have worried that ozone holes might start appearing over other areas of the globe, though to date the only other large-scale depletion is a smaller ozone "dimple" observed during the Arctic spring around the North Pole. Ozone at middle latitudes has declined, but by a much smaller extent (about 4–5 percent decrease).

- If stratospheric conditions become more severe (cooler temperatures, more clouds, more active chlorine), global ozone may decrease at a greater pace. Standard global warming theory predicts that the stratosphere will cool.

- When the Antarctic ozone hole breaks up each year, the ozone-depleted air drifts out into nearby regions. Decreases in the ozone level of up to 10 percent have been reported in New Zealand in the month following the breakup of the Antarctic ozone hole, with ultraviolet-B radiation intensities increasing by more than 15 percent since the 1970s.

Consequences of Ozone Layer Depletion

Since the ozone layer absorbs UVB ultraviolet light from the sun, ozone layer depletion increases surface UVB levels (all else equal), which could lead to damage, including increase in skin cancer. This was the reason for the Montreal Protocol. Although decreases in stratospheric ozone are well-tied to CFCs and to increases in surface UVB, there is no direct observational evidence linking ozone depletion to higher incidence of skin cancer and eye damage in human beings. This is partly because UVA, which has also been implicated in some forms of skin cancer, is not absorbed by ozone, and because it is nearly impossible to control statistics for lifestyle changes over time.

Increased UV

Ozone, while a minority constituent in Earth's atmosphere, is responsible for most of the absorption of UVB radiation. The amount of UVB radiation that penetrates through the ozone layer decreases exponentially with the slant-path thickness and density of the layer. When stratospheric ozone levels decrease, higher levels of UVB reach the Earth's surface. UV-driven phenolic formation in tree rings has dated the start of ozone depletion in northern latitudes to the late 1700s.

In October 2008, the Ecuadorian Space Agency published a report called HIPERION, a study of the last 28 years data from 10 satellites and dozens of ground instruments around the world among them their own, and found that the UV radiation reaching equatorial latitudes was far greater than expected, with the UV Index climbing as high as 24 in some very populated cities; the WHO considers 11 as an extreme index and a great risk to health. The report concluded that depleted ozone levels around the mid-latitudes of the planet are already endangering large populations in these areas. Later, the CONIDA, the Peruvian Space Agency, published its own study, which yielded almost the same findings as the Ecuadorian study.

Biological Effects

The main public concern regarding the ozone hole has been the effects of increased surface UV radiation on human health. So far, ozone depletion in most locations has been typically a few percent and, as noted above, no direct evidence of health damage is available in most latitudes. If the high levels of depletion seen in the ozone hole were to be common across the globe, the effects could be substantially more dramatic. As the ozone hole over Antarctica has in some instances grown so large as to affect parts of Australia, New Zealand, Chile, Argentina, and South Africa, environmentalists have been concerned that the increase in surface UV could be significant.

Ozone depletion would magnify all of the effects of UV on human health, both positive (including production of Vitamin D) and negative (including sunburn, skin cancer, and cataracts). In addition, increased surface UV leads to increased tropospheric ozone, which is a health risk to humans.

Basal and Squamous Cell Carcinomas

The most common forms of skin cancer in humans, basal and squamous cell carcinomas, have been strongly linked to UVB exposure. The mechanism by which UVB induces these cancers is well understood—absorption of UVB radiation causes the pyrimidine bases in the DNA molecule to form dimers, resulting in transcription errors when the DNA replicates. These cancers are relatively mild and rarely fatal, although the treatment of squamous cell carcinoma sometimes requires extensive reconstructive surgery. By combining epidemiological data with results of animal studies, scientists have estimated that every one percent decrease in long-term stratospheric ozone would increase the incidence of these cancers by two percent.

Malignant Melanoma

Another form of skin cancer, malignant melanoma, is much less common but far more dangerous, being lethal in about 15–20 percent of the cases diagnosed. The relationship between malignant melanoma and ultraviolet exposure is not yet fully understood, but it appears that both UVB and UVA are involved. Because of this uncertainty, it is difficult to estimate the effect of ozone depletion on melanoma incidence. One study showed that a 10 percent increase in UVB radiation was associated with a 19 percent increase in melanomas for men and 16 percent for women. A study of people in Punta Arenas, at the southern tip of Chile, showed a 56 percent increase in melanoma and a 46 percent increase in nonmelanoma skin cancer over a period of seven years, along with decreased ozone and increased UVB levels.

Cortical Cataracts

Epidemiological studies suggest an association between ocular cortical cataracts and UVB exposure, using crude approximations of exposure and various cataract assessment techniques. A detailed assessment of ocular exposure to UVB was carried out in a study on Chesapeake Bay Watermen, where increases in average annual ocular exposure were associated with increasing risk of cortical opacity. In this highly exposed group of predominantly white males, the evidence linking cortical opacities to sunlight exposure was the strongest to date. Based on these results, ozone depletion is predicted to cause hundreds of thousands of additional cataracts by 2050.

Increased Tropospheric Ozone

Increased surface UV leads to increased tropospheric ozone. Ground-level ozone is generally recognized to be a health risk, as ozone is toxic due to its strong oxidant properties. The risks are particularly high for young children, the elderly, and those with asthma or other respiratory difficulties. At this time, ozone at ground level is produced mainly by the action of UV radiation on combustion gases from vehicle exhausts.

Increased Production of Vitamin D

Vitamin D is produced in the skin by ultraviolet light. Thus, higher UVB exposure raises human vitamin D in those deficient in it. Recent research (primarily since the Montreal Protocol) shows that many humans have less than optimal vitamin D levels. In particular, in the U.S. population, the lowest quarter of vitamin D (<17.8 ng/ml) were found using information from the National Health and Nutrition Examination Survey to be associated with an increase in all-cause mortality in the general population. While blood level of Vitamin D in excess of 100 ng/ml appear to raise blood calcium excessively and to be associated with higher mortality, the body has mechanisms that prevent sunlight from producing Vitamin D in excess of the body's requirements.

Effects on Non-human Animals

A November 2010 report by scientists at the Institute of Zoology in London found that whales off the coast of California have shown a sharp rise in sun damage, and these scientists "fear that the thinning ozone layer is to blame". The study photographed and took skin biopsies from over 150 whales in the Gulf of California and found "widespread evidence of epidermal damage commonly associated with acute and severe sunburn", having cells that form when the DNA is damaged by UV radiation. The findings suggest "rising UV levels as a result of ozone depletion are to blame for the observed skin damage, in the same way that human skin cancer rates have been on the increase in recent decades."

Effects on Crops

An increase of UV radiation would be expected to affect crops. A number of economically important species of plants, such as rice, depend on cyanobacteria residing on their roots for the retention of nitrogen. Cyanobacteria are sensitive to UV radiation and would be affected by its increase. "Despite mechanisms to reduce or repair the effects of increased ultraviolet radiation, plants have a limited ability to adapt to increased levels of UVB, therefore plant growth can be directly affected by UVB radiation."

Public Policy

The full extent of the damage that CFCs have caused to the ozone layer is not known and will not be known for decades; however, marked decreases in column ozone have already been observed. The Montreal and Vienna conventions were installed long before a scientific consensus was established or important uncertainties in the science field were being resolved. The ozone case was understood comparably well by lay persons as e.g. *Ozone shield* or *ozone hole* were useful "easy-to-understand bridging metaphors". Americans voluntarily switched away from aerosol sprays, resulting in a 50 percent sales loss even before legislation was enforced.

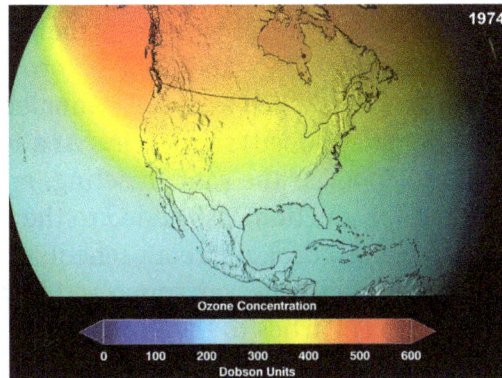

NASA projections of stratospheric ozone concentrations if chlorofluorocarbons had not been banned

After a 1976 report by the United States National Academy of Sciences concluded that credible scientific evidence supported the ozone depletion hypothesis a few countries, including the United States, Canada, Sweden, Denmark, and Norway, moved to eliminate the use of CFCs in aerosol spray cans. At the time this was widely regarded as a first step towards a more comprehensive regulation policy, but progress in this direction slowed in subsequent years, due to a combination of political factors (continued resistance from the halocarbon industry and a general change in attitude towards environmental regulation during the first two years of the Reagan administration) and scientific developments (subsequent National Academy assessments that indicated that the first estimates of the magnitude of ozone depletion had been overly large). A critical DuPont manufacturing patent for Freon was set to expire in 1979. The United States banned the use of CFCs in aerosol cans in 1978. The European Community rejected proposals to ban CFCs in aerosol sprays, and in the U.S., CFCs continued to be used as refrigerants and for cleaning circuit boards. Worldwide CFC production fell sharply after the U.S. aerosol ban, but by 1986 had returned nearly to its 1976 level. In 1993, DuPont shut down its CFC facility.

The U.S. Government's attitude began to change again in 1983, when William Ruckelshaus replaced Anne M. Burford as Administrator of the United States Environmental Protection Agency. Under Ruckelshaus and his successor, Lee Thomas, the EPA pushed for an international approach to halocarbon regulations. In 1985 20 nations, including most of the major CFC producers, signed the Vienna Convention for the Protection of the Ozone Layer, which established a framework for negotiating international regulations on ozone-depleting substances. That same year, the discovery of the Antarctic ozone hole was announced, causing a revival in public attention to the issue. In 1987, representatives from 43 nations signed the Montreal Protocol. Meanwhile, the halocarbon industry shifted its position and started supporting a protocol to limit CFC production. However, this shift was uneven with DuPont acting more quickly than their European counterparts. DuPont may have feared court action related to increased skin cancer especially as the EPA had published a study in 1986 claiming that an additional 40 million cases and 800,000 cancer deaths were to be expected in the U.S. in the next 88 years. The EU shifted its position as well after Germany gave up its defence of the CFC industry and started supporting moves towards regulation. Government and industry in France and the UK tried to defend their CFC producing industries even after the Montreal Protocol had been signed.

At Montreal, the participants agreed to freeze production of CFCs at 1986 levels and to reduce production by 50 percent by 1999. After a series of scientific expeditions to the Antarctic produced convincing evidence that the ozone hole was indeed caused by chlorine and bromine from manmade organohalogens, the Montreal Protocol was strengthened at a 1990 meeting in London. The participants agreed to phase out CFCs and halons entirely (aside from a very small amount marked for certain "essential" uses, such as asthma inhalers) by 2000 in non-Article 5 countries and by 2010 in Article 5 (less developed) signatories. At a 1992 meeting in Copenhagen, the phase-out date was moved up to 1996. At the same meeting, methyl bromide (MeBr), a fumigant used primarily in agricultural production, was added to the list of controlled substances. For all substances controlled under the protocol, phaseout schedules were delayed for less developed countries, and phaseout in these countries was supported by transfers of expertise, technology, and money from non-Article 5(1) Parties to the Protocol. Additionally, exemptions from the agreed schedules could be applied for under the Essential Use Exemption (EUE) process for substances other than methyl bromide and under the Critical Use Exemption (CUE) process for methyl bromide.

A hydrocarbon refrigerant was developed in 1992 at the request of the non-governmental organization (NGO) Greenpeace, and was being used by some 40 percent of the refrigerator market in 2013. In the U.S., however, change has been much slower. To some extent, CFCs were being replaced by the less damaging hydrochlorofluorocarbons (HCFCs), although concerns remain regarding HCFCs also. In some applications, hydrofluorocarbons (HFCs) were being used to replace CFCs. HFCs, which contain no chlorine or bromine, do not contribute at all to ozone depletion although they are potent greenhouse gases. The best known of these compounds is probably HFC-134a (R-134a), which in the United States has largely replaced CFC-12 (R-12) in automobile air conditioners. In laboratory analytics (a former "essential" use) the ozone depleting substances can be replaced with various other solvents. The development and promotion of an ozone-safe hydrocarbon refrigerant was a breakthrough which occurred after the initiative of a non-governmental organization (NGO). Civil society including especially NGOs, in fact, played critical roles at all stages of policy development leading up to the Vienna Conference, the Montreal Protocol, and in assessing compliance afterwards. The major companies claimed that no alternatives to HFC existed. By 1992, Greenpeace had requested a scientific team to research an ozone-safe refrigerant, an effort which resulted in a successful hydrocarbon formula, for which Greenpeace left the patent as open source. The NGO maintained the technology as an open patent as it worked successfully first with a small company to market an appliance beginning in Europe, then Asia and later Latin America, receiving a 1997 UNEP award. By 1995, Germany had already made CFC refrigerators illegal. Production spread to companies like Electrolux, Bosch, and LG, with sales reaching some 300 million refrigerators by 2008. However, the giant corporations all continued to refuse to apply the technology in Latin America. In 2003, a domestic Argentinian company began Greenfreeze production, while the giant Bosch in Brazil began a year later. Chemical companies like Du Pont, whose representatives even disparaged the green technology as "that German technology," maneuvered the EPA to block the technology in the U.S. until 2011. Ben & Jerry's of Unilever and General Electric, spurred by Greenpeace, had expressed formal interest in 2008 which figured in the EPA's final approval. Currently, more than 600 million refrigerators have been sold, making up more than 40 percent of the market. Since 2004, corporations like Coca-Cola, Carlsberg, and IKEA have been forming a coalition to promote the ozone-safe Greenfreeze units.

More recently, policy experts have advocated for efforts to link ozone protection efforts to climate protection efforts. Many ODS are also greenhouse gases, some thousands of times more powerful agents of radiative forcing than carbon dioxide over the short and medium term. Thus policies protecting the ozone layer have had benefits in mitigating climate change. In fact, the reduction of the radiative forcing due to ODS probably masked the true level of climate change effects of other GHGs, and was responsible for the "slow down" of global warming from the mid-90s. Policy decisions in one arena affect the costs and effectiveness of environmental improvements in the other.

ODS Requirements in the Marine Industry

The IMO has amended MARPOL Annex VI Regulation 12 regarding ozone depleting substances. As from July 1, 2010, all vessels where MARPOL Annex VI is applicable should have a list of equipment using ozone depleting substances. The list should include name of ODS, type and location of equipment, quantity in kg and date. All changes since that date should be recorded in an ODS Record book on board recording all intended or unintended releases to the atmosphere. Furthermore, new ODS supply or landing to shore facilities should be recorded as well.

Prospects of Ozone Depletion

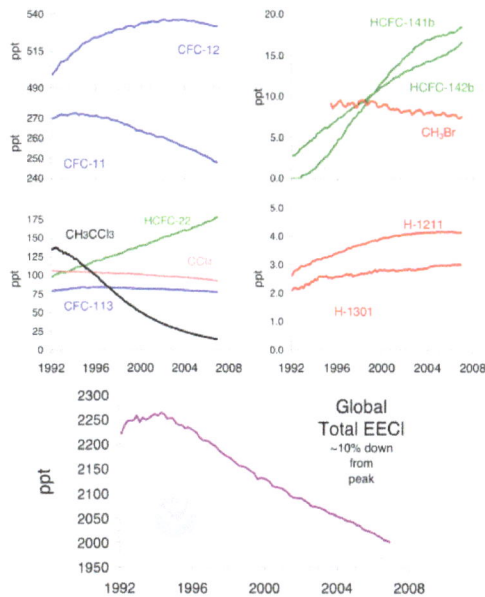

Ozone-depleting gas trends

Since the adoption and strengthening of the Montreal Protocol has led to reductions in the emissions of CFCs, atmospheric concentrations of the most-significant compounds have been declining. These substances are being gradually removed from the atmosphere; since peaking in 1994, the Effective Equivalent Chlorine (EECl) level in the atmosphere had dropped about 10 percent by 2008. The decrease in ozone-depleting chemicals has also been significantly affected by a decrease in bromine-containing chemicals. The data suggest that substantial natural sources exist for atmospheric methyl bromide (C3Br). The phase-out of CFCs means that nitrous oxide (N2O), which is not covered by the Montreal Protocol, has become the most highly emitted ozone-depleting substance and is expected to remain so throughout the 21st century.

A 2005 IPCC review of ozone observations and model calculations concluded that the global amount of ozone has now approximately stabilized. Although considerable variability is expected from year to year, including in polar regions where depletion is largest, the ozone layer is expected to begin to recover in coming decades due to declining ozone-depleting substance concentrations, assuming full compliance with the Montreal Protocol.

The Antarctic ozone hole is expected to continue for decades. Ozone concentrations in the lower stratosphere over Antarctica will increase by 5–10 percent by 2020 and return to pre-1980 levels by about 2060–2075. This is 10–25 years later than predicted in earlier assessments, because of revised estimates of atmospheric concentrations of ozone-depleting substances, including a larger predicted future usage in developing countries. Another factor that may prolong ozone depletion is the drawdown of nitrogen oxides from above the stratosphere due to changing wind patterns. A gradual trend toward "healing" was reported in 2016.

Research History

The basic physical and chemical processes that lead to the formation of an ozone layer in the Earth's stratosphere were discovered by Sydney Chapman in 1930. Short-wavelength UV radiation splits an oxygen (O_2) molecule into two oxygen (O) atoms, which then combine with other oxygen molecules to form ozone. Ozone is removed when an oxygen atom and an ozone molecule "recombine" to form two oxygen molecules, i.e. $O + O_3 \rightarrow 2O_2$. In the 1950s, David Bates and Marcel Nicolet presented evidence that various free radicals, in particular hydroxyl (OH) and nitric oxide (NO), could catalyze this recombination reaction, reducing the overall amount of ozone. These free radicals were known to be present in the stratosphere, and so were regarded as part of the natural balance—it was estimated that in their absence, the ozone layer would be about twice as thick as it currently is.

In 1970 Paul Crutzen pointed out that emissions of nitrous oxide (N_2O), a stable, long-lived gas produced by soil bacteria, from the Earth's surface could affect the amount of nitric oxide (NO) in the stratosphere. Crutzen showed that nitrous oxide lives long enough to reach the stratosphere, where it is converted into NO. Crutzen then noted that increasing use of fertilizers might have led to an increase in nitrous oxide emissions over the natural background, which would in turn result in an increase in the amount of NO in the stratosphere. Thus human activity could affect the stratospheric ozone layer. In the following year, Crutzen and (independently) Harold Johnston suggested that NO emissions from supersonic passenger aircraft, which would fly in the lower stratosphere, could also deplete the ozone layer. However, more recent analysis in 1995 by David W. Fahey, an atmospheric scientist at the National Oceanic and Atmospheric Administration, found that the drop in ozone would be from 1–2 percent if a fleet of 500 supersonic passenger aircraft were operated. This, Fahey expressed, would not be a showstopper for advanced supersonic passenger aircraft development.

Rowland–Molina Hypothesis

In 1974 Frank Sherwood Rowland, Chemistry Professor at the University of California at Irvine, and his postdoctoral associate Mario J. Molina suggested that long-lived organic halogen compounds, such as CFCs, might behave in a similar fashion as Crutzen had proposed for nitrous oxide. James Lovelock had recently discovered, during a cruise in the South Atlantic in 1971, that almost all of the CFC compounds manufactured since their invention in 1930 were still present in the atmosphere. Molina and Rowland concluded that, like N2O, the CFCs would reach the stratosphere where they would be dissociated by UV light, releasing chlorine atoms. A year earlier, Richard Stolarski and Ralph Cicerone at the University of Michigan had shown that Cl is even more efficient than NO at catalyzing the destruction of ozone. Similar conclusions were reached by Michael McElroy and Steven Wofsy at Harvard University. Neither group, however, had realized that CFCs were a potentially large source of stratospheric chlorine—instead, they had been investigating the possible effects of HCl emissions from the Space Shuttle, which are very much smaller.

The Rowland–Molina hypothesis was strongly disputed by representatives of the aerosol and halocarbon industries. The Chair of the Board of DuPont was quoted as saying that ozone depletion theory is "a science fiction tale ... a load of rubbish ... utter nonsense". Robert Abplanalp, the President of Precision Valve Corporation (and inventor of the first practical aerosol spray can valve), wrote to the Chancellor of UC Irvine to complain about Rowland's public statements. Nevertheless, within three years most of the basic assumptions made by Rowland and Molina were confirmed by laboratory measurements and by direct observation in the stratosphere. The concentrations of the source gases (CFCs and related compounds) and the chlorine reservoir species (HCl and ClONO2) were measured throughout the stratosphere, and demonstrated that CFCs were indeed the major source of stratospheric chlorine, and that nearly all of the CFCs emitted would eventually reach the stratosphere. Even more convincing was the measurement, by James G. Anderson and collaborators, of chlorine monoxide (ClO) in the stratosphere. ClO is produced by the reaction of Cl with ozone—its observation thus demonstrated that Cl radicals not only were present in the stratosphere but also were actually involved in destroying ozone. McElroy and Wofsy extended the work of Rowland and Molina by showing that bromine atoms were even more effective catalysts for ozone loss than chlorine atoms and argued that the brominated organic compounds known as halons, widely used in fire extinguishers, were a potentially large source of stratospheric bromine. In 1976 the United States National Academy of Sciences released a report concluding that the ozone depletion hypothesis was strongly supported by the scientific evidence. Scientists calculated that if CFC production continued to increase at the going rate of 10 percent per year until 1990 and then remain steady, CFCs would cause a global ozone loss of 5–7 percent by 1995, and a 30–50 percent loss by 2050. In response the United States, Canada and Norway banned the use of CFCs in aerosol spray cans in 1978. However, subsequent research, summarized by the National Academy in reports issued between 1979 and 1984, appeared to show that the earlier estimates of global ozone loss had been too large.

Crutzen, Molina, and Rowland were awarded the 1995 Nobel Prize in Chemistry for their work on stratospheric ozone.

Antarctic Ozone Hole

The discovery of the Antarctic "ozone hole" by British Antarctic Survey scientists Farman, Gardiner and Shanklin (first reported in a paper in *Nature* in May 1985) came as a shock to the scientific community, because the observed decline in polar ozone was far larger than anyone had anticipated. Satellite measurements showing massive depletion of ozone around the south pole were becoming available at the same time. However, these were initially rejected as unreasonable by data quality control algorithms (they were filtered out as errors since the values were unexpectedly low); the ozone hole was detected only in satellite data when the raw data was reprocessed following evidence of ozone depletion in *in situ* observations. When the software was rerun without the flags, the ozone hole was seen as far back as 1976.

Susan Solomon, an atmospheric chemist at the National Oceanic and Atmospheric Administration (NOAA), proposed that chemical reactions on polar stratospheric clouds (PSCs) in the cold Antarctic stratosphere caused a massive, though localized and seasonal, increase in the amount of chlorine present in active, ozone-destroying forms. The polar stratospheric clouds in Antarctica are only formed when there are very low temperatures, as low as −80 °C, and early spring conditions. In such conditions the ice crystals of the cloud provide a suitable surface for conversion of unreactive chlorine compounds into reactive chlorine compounds, which can deplete ozone easily.

Moreover, the polar vortex formed over Antarctica is very tight and the reaction occurring on the surface of the cloud crystals is far different from when it occurs in atmosphere. These conditions have led to ozone hole formation in Antarctica. This hypothesis was decisively confirmed, first by laboratory measurements and subsequently by direct measurements, from the ground and from high-altitude airplanes, of very high concentrations of chlorine monoxide (ClO) in the Antarctic stratosphere.

Alternative hypotheses, which had attributed the ozone hole to variations in solar UV radiation or to changes in atmospheric circulation patterns, were also tested and shown to be untenable.

Meanwhile, analysis of ozone measurements from the worldwide network of ground-based Dobson spectrophotometers led an international panel to conclude that the ozone layer was in fact being depleted, at all latitudes outside of the tropics. These trends were confirmed by satellite measurements. As a consequence, the major halocarbon-producing nations agreed to phase out production of CFCs, halons, and related compounds, a process that was completed in 1996.

Since 1981 the United Nations Environment Programme, under the auspices of the World Meteorological Organization, has sponsored a series of technical reports on the Scientific Assessment of Ozone Depletion, based on satellite measurements. The 2007 report showed that the hole in the ozone layer was recovering and the smallest it had been for about a decade. The 2010 report found, "Over the past decade, global ozone and ozone in the Arctic and Antarctic regions is no longer decreasing but is not yet increasing. The ozone layer outside the Polar regions is projected to recover to its pre-1980 levels some time before the middle of this century. In contrast, the springtime ozone hole over the Antarctic is expected to recover much later." In 2012, NOAA and NASA reported "Warmer air temperatures high above the Antarctic led to the second smallest season ozone hole in 20 years averaging 17.9 million square kilometres. The hole reached its maximum size for the season on Sept 22, stretching to 21.2 million square kilometres." A gradual trend toward "healing" was reported in 2016.

The hole in the Earth's ozone layer over the South Pole has affected atmospheric circulation in the Southern Hemisphere all the way to the equator. The ozone hole has influenced atmospheric circulation all the way to the tropics and increased rainfall at low, subtropical latitudes in the Southern Hemisphere.

Arctic Ozone Hole

On March 15, 2011, a record ozone layer loss was observed, with about half of the ozone present over the Arctic having been destroyed. The change was attributed to increasingly cold winters in the Arctic stratosphere at an altitude of approximately 20 km (12 mi), a change associated with global warming in a relationship that is still under investigation. By March 25, the ozone loss had become the largest compared to that observed in all previous winters with the possibility that it would become an ozone hole. This would require that the quantities of ozone to fall below 200 Dobson units, from the 250 recorded over central Siberia. It is predicted that the thinning layer would affect parts of Scandinavia and Eastern Europe on March 30–31.

On March 3, 2005, the journal Nature published an article linking 2004's unusually large Arctic ozone hole to solar wind activity.

On October 2, 2011, a study was published in the journal *Nature*, which said that between December 2010 and March 2011 up to 80 percent of the ozone in the atmosphere at about 20 kilometres (12 mi) above the surface was destroyed. The level of ozone depletion was severe enough that scientists said it could be compared to the ozone hole that forms over Antarctica every winter. According to the study, "for the first time, sufficient loss occurred to reasonably be described as an Arctic ozone hole." The study analyzed data from the Aura and CALIPSO satellites, and determined that the larger-than-normal ozone loss was due to an unusually long period of cold weather in the Arctic, some 30 days more than typical, which allowed for more ozone-destroying chlorine compounds to be created. According to Lamont Poole, a co-author of the study, cloud and aerosol particles on which the chlorine compounds are found "were abundant in the Arctic until mid March 2011—much later than usual—with average amounts at some altitudes similar to those observed in the Antarctic, and dramatically larger than the near-zero values seen in March in most Arctic winters".

Tibet Ozone Hole

As winters that are colder are more affected, at times there is an ozone hole over Tibet. In 2006, a 2.5 million square kilometer ozone hole was detected over Tibet. Also again in 2011 an ozone hole appeared over mountainous regions of Tibet, Xinjiang, Qinghai and the Hindu Kush, along with an unprecedented hole over the Arctic, though the Tibet one is far less intense than the ones over the Arctic or Antarctic.

Potential Depletion by Storm Clouds

Research in 2012 showed that the same process that produces the ozone hole over Antarctica occurs over summer storm clouds in the United States, and thus may be destroying ozone there as well.

Ozone Depletion and Global Warming

Among others, Robert Watson had a role in the science assessment and in the regulation efforts of ozone depletion and global warming. Prior to the 1980s, the EU, NASA, NAS, UNEP, WMO and the British government had dissenting scientific reports and Watson played a crucial role in the process of unified assessments. Based on the experience with the ozone case, the IPCC started to work on a unified reporting and science assessment to reach a consensus to provide the IPCC Summary for Policymakers.

There are various areas of linkage between ozone depletion and global warming science:

Radiative forcing components

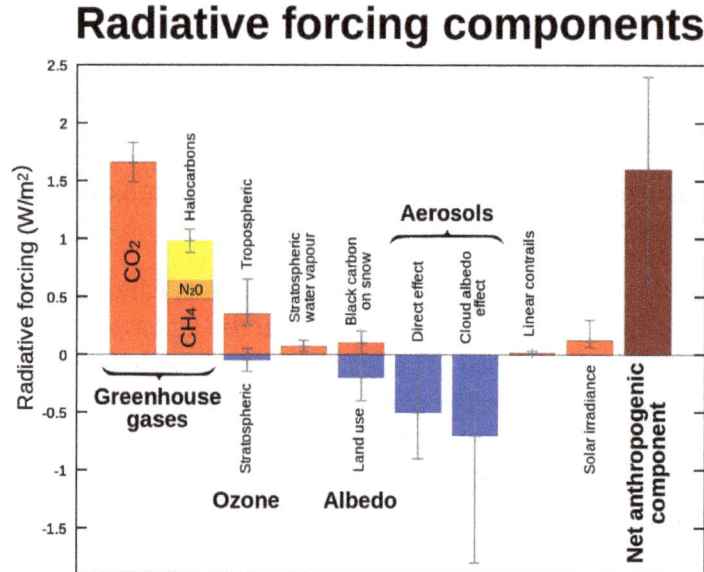

Radiative forcing from various greenhouse gases and other sources

- The same CO_2 radiative forcing that produces global warming is expected to cool the stratosphere. This cooling, in turn, is expected to produce a relative *increase* in ozone (O_3) depletion in polar area and the frequency of ozone holes.

- Conversely, ozone depletion represents a radiative forcing of the climate system. There are two opposing effects: Reduced ozone causes the stratosphere to absorb less solar radiation, thus cooling the stratosphere while warming the troposphere; the resulting colder stratosphere emits less long-wave radiation downward, thus cooling the troposphere. Overall, the cooling dominates; the IPCC concludes *"observed stratospheric O_3 losses over the past two decades have caused a negative forcing of the surface-troposphere system"* of about -0.15 ± 0.10 watts per square meter (W/m^2).

- One of the strongest predictions of the greenhouse effect is that the stratosphere will cool. Although this cooling has been observed, it is not trivial to separate the effects of changes in the concentration of greenhouse gases and ozone depletion since both will lead to cooling. However, this can be done by numerical stratospheric modeling. Results from the National Oceanic and Atmospheric Administration's Geophysical Fluid Dynamics Laboratory show that above 20 km (12 mi), the greenhouse gases dominate the cooling.

- As noted under 'Public Policy', ozone depleting chemicals are also often greenhouse gases. The increases in concentrations of these chemicals have produced 0.34 ± 0.03 W/m^2 of radiative forcing, corresponding to about 14 percent of the total radiative forcing from increases in the concentrations of well-mixed greenhouse gases.

- The long term modeling of the process, its measurement, study, design of theories and testing take decades to document, gain wide acceptance, and ultimately become the dominant paradigm. Several theories about the destruction of ozone were hypothesized in the 1980s, published in the late 1990s, and are currently being investigated. Dr Drew Schindell, and Dr Paul Newman, NASA Goddard, proposed a theory in the late 1990s, using computational modeling methods to model ozone destruction, that accounted for 78 percent of the ozone destroyed. Further refinement of that model accounted for 89 percent of the ozone destroyed, but pushed back the estimated recovery of the ozone hole from 75 years to 150 years. (An important part of that model is the lack of stratospheric flight due to depletion of fossil fuels.)

Misconceptions

CFC Weight

Since CFC molecules are heavier than air (nitrogen or oxygen), it is commonly believed that the CFC molecules cannot reach the stratosphere in significant amount. However, atmospheric gases are not sorted by weight; the forces of wind can fully mix the gases in the atmosphere. Lighter CFCs are evenly distributed throughout the turbosphere and reach the upper atmosphere, although some of the heavier CFCs are not evenly distributed.

Percentage of Man-made Chlorine

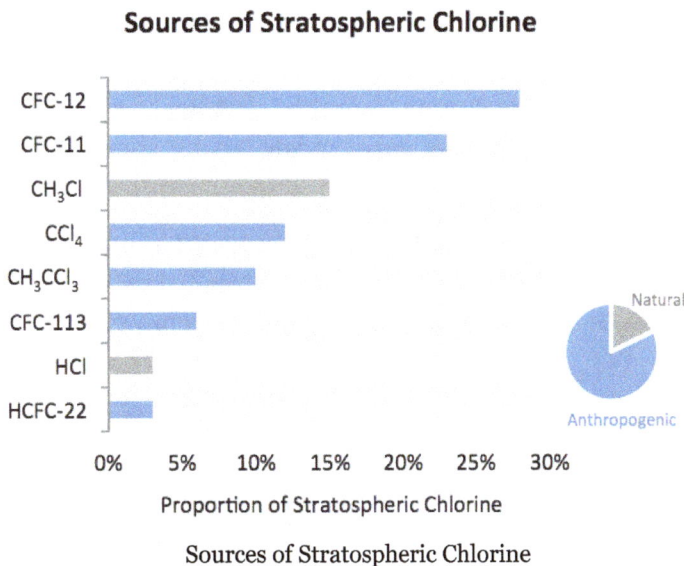

Sources of Stratospheric Chlorine

Sources of Stratospheric Chlorine

Another misconception is that "it is generally accepted that natural sources of tropospheric chlorine are four to five times larger than man-made ones." While strictly true, *tropospheric* chlorine is irrelevant; it is *stratospheric* chlorine that affects ozone depletion. Chlorine from ocean spray is soluble and thus is washed by rainfall before it reaches the stratosphere. CFCs, in contrast, are insoluble and long-lived, allowing them to reach the stratosphere. In the lower atmosphere, there is much more chlorine from CFCs and related haloalkanes than there is in HCl from salt spray, and in the stratosphere halocarbons are dominant. Only methyl chloride, which is one of these halocarbons, has a mainly natural source, and it is responsible for about 20 percent of the chlorine in the stratosphere; the remaining 80 percent comes from manmade sources.

Very violent volcanic eruptions can inject HCl into the stratosphere, but researchers have shown that the contribution is not significant compared to that from CFCs. A similar erroneous assertion is that soluble halogen compounds from the volcanic plume of Mount Erebus on Ross Island, Antarctica are a major contributor to the Antarctic ozone hole.

Nevertheless, the latest study showed that the role of Mount Erebus volcano in the Antarctic ozone depletion was probably underestimated. Based on the NCEP/NCAR reanalysis data over the last 35 years and by using the NOAA HYSPLIT trajectory model, researchers showed that Erebus volcano gas emissions (including hydrogen chloride (HCl)) can reach the Antarctic stratosphere via high-latitude cyclones and then the polar vortex. Depending on Erebus volcano activity, the additional annual HCl mass entering the stratosphere from Erebus varies from 1.0 to 14.3 kt.

First Observation

G.M.B. Dobson (Exploring the Atmosphere, 2nd Edition, Oxford, 1968) mentioned that when springtime ozone levels in the Antarctic over Halley Bay were first measured in 1956, he was surprised to find that they were ~320 DU, or about 150 DU below spring Arctic levels of ~450 DU. These were at that time the only known Antarctic ozone values available. What Dobson describes is essentially the *baseline* from which the ozone hole is measured: actual ozone hole values are in the 150–100 DU range.

The discrepancy between the Arctic and Antarctic noted by Dobson was primarily a matter of timing: during the Arctic spring ozone levels rose smoothly, peaking in April, whereas in the Antarctic they stayed approximately constant during early spring, rising abruptly in November when the polar vortex broke down.

The behavior seen in the Antarctic ozone hole is completely different. Instead of staying constant, early springtime ozone levels suddenly drop from their already low winter values, by as much as 50 percent, and normal values are not reached again until December.

Location of Hole

Some people thought that the ozone hole should be above the sources of CFCs. However, CFCs are well mixed globally in the troposphere and stratosphere. The reason for occurrence of the ozone hole above Antarctica is not because there are more CFCs concentrated but because the low temperatures help form polar stratospheric clouds. In fact, there are findings of significant and localized "ozone holes" above other parts of the earth.

World Ozone Day

In 1994, the United Nations General Assembly voted to designate September 16 as "World Ozone Day", to commemorate the signing of the Montreal Protocol on that date in 1987.

Ocean Acidification

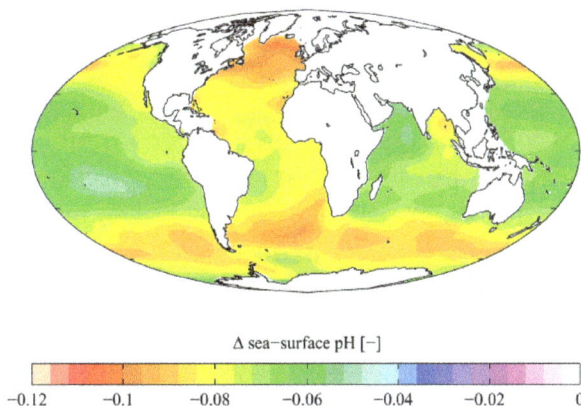

Δ sea−surface pH [−]

Estimated change in sea water pH caused by human created CO2 between the 1700s and the 1990s, from the Global Ocean Data Analysis Project (GLODAP) and the World Ocean Atlas

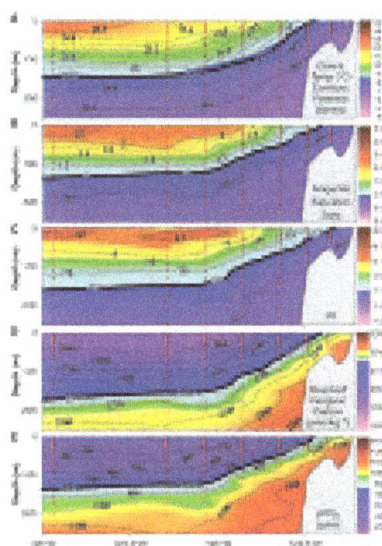

NOAA provides evidence for upwelling of "acidified" water onto the Continental Shelf. In the figure above, note the vertical sections of (A) temperature, (B) aragonite saturation, (C) pH, (D) DIC, and (E) pCO2 on transect line 5 off Pt. St. George, California. The potential density surfaces are superimposed on the temperature section. The 26.2 potential density surface delineates the location of the first instance in which the undersaturated water is upwelled from depths of 150 to 200 m onto the shelf and outcropping at the surface near the coast. The red dots represent sample locations.

Ocean acidification is the ongoing decrease in the pH of the Earth's oceans, caused by the uptake of carbon dioxide (CO_2) from the atmosphere. Seawater is slightly basic (meaning pH > 7), and the process in question is a shift towards pH-neutral conditions rather than a transition to acidic conditions (pH < 7). Ocean alkalinity is not changed by the process, or may increase over long time periods due to carbonate dissolution. An estimated 30–40% of the carbon dioxide from human activity released into the atmosphere dissolves into oceans, rivers and lakes. To achieve chemical equilibrium, some of it reacts with the water to form carbonic acid. Some of these extra carbonic acid molecules react with a water molecule to give a bicarbonate ion and a hydronium ion, thus increasing ocean acidity (H⁺ ion concentration). Between 1751 and 1994 surface ocean pH is estimated to have decreased from approximately 8.25 to 8.14, representing an increase of almost 30% in H^+ ion concentration in the world's oceans. Earth System Models project that within the last decade ocean acidity exceeded historical analogs and in combination with other ocean biogeochemical changes could undermine the functioning of marine ecosystems and disrupt the provision of many goods and services associated with the ocean.

Increasing acidity is thought to have a range of potentially harmful consequences for marine organisms, such as depressing metabolic rates and immune responses in some organisms, and causing coral bleaching. By increasing the presence of free hydrogen ions, each molecule of carbonic acid that forms in the oceans ultimately results in the conversion of *two* carbonate ions into bicarbonate ions. This net decrease in the amount of carbonate ions available makes it more difficult for marine calcifying organisms, such as coral and some plankton, to form biogenic calcium carbonate, and such structures become vulnerable to dissolution. Ongoing acidification of the oceans threatens food chains connected with the oceans. As members of the InterAcademy Panel, 105 science academies have issued a statement on ocean acidification recommending that by 2050, global CO_2 emissions be reduced by at least 50% compared to the 1990 level.

While ongoing ocean acidification is anthropogenic in origin, it has occurred previously in Earth's history. The most notable example is the Paleocene-Eocene Thermal Maximum (PETM), which occurred approximately 56 million years ago. For reasons that are currently uncertain, massive amounts of carbon entered the ocean and atmosphere, and led to the dissolution of carbonate sediments in all ocean basins.

Ocean acidification has been called the "evil twin of global warming" and "the other CO_2 problem".

Carbon Cycle

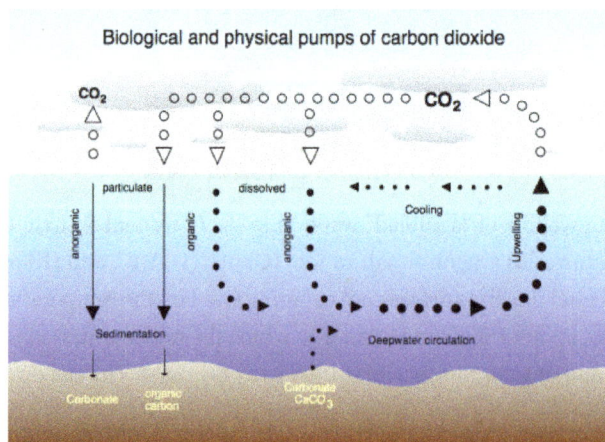

The CO2 cycle between the atmosphere and the ocean

The carbon cycle describes the fluxes of carbon dioxide (CO2) between the oceans, terrestrial biosphere, lithosphere, and the atmosphere. Human activities such as the combustion of fossil fuels and land use changes have led to a new flux of CO2 into the atmosphere. About 45% has remained in the atmosphere; most of the rest has been taken up by the oceans, with some taken up by terrestrial plants.

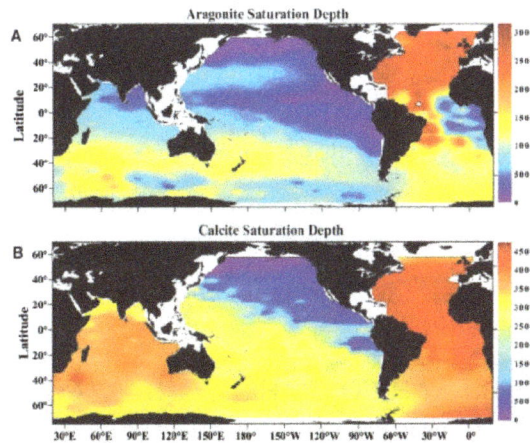

Distribution of (A) aragonite and (B) calcite saturation depth in the global oceans

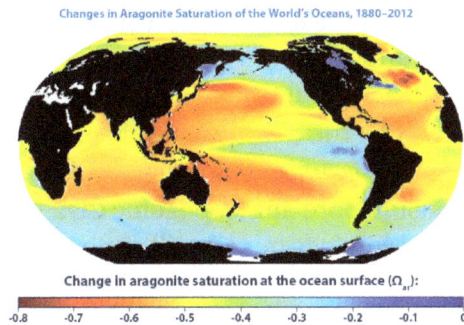

The map was created by the National Oceanic and Atmospheric Administration and the Woods Hole Oceanographic Institution using Community Earth System Model data. This map was created by comparing average conditions during the 1880s with average conditions during the most recent 10 years (2003–2012). Aragonite saturation has only been measured at selected locations during the last few decades, but it can be calculated reliably for different times and locations based on the relationships scientists have observed among aragonite saturation, pH, dissolved carbon, water temperature, concentrations of carbon dioxide in the atmosphere, and other factors that can be measured. This map shows changes in the amount of aragonite dissolved in ocean surface waters between the 1880s and the most recent decade (2003–2012). Aragonite saturation is a ratio that compares the amount of aragonite that is actually present with the total amount of aragonite that the water could hold if it were completely saturated. The more negative the change in aragonite saturation, the larger the decrease in aragonite available in the water, and the harder it is for marine creatures to produce their skeletons and shells. The global map shows changes over time in the amount of aragonite dissolved in ocean water, which is called aragonite saturation.

The carbon cycle involves both organic compounds such as cellulose and inorganic carbon compounds such as carbon dioxide and the carbonates. The inorganic compounds are particularly relevant when discussing ocean acidification for it includes many forms of dissolved CO_2 present in the Earth's oceans.

When CO_2 dissolves, it reacts with water to form a balance of ionic and non-ionic chemical species: dissolved free carbon dioxide (CO_2(aq)), carbonic acid (H_2CO_3), bicarbonate (HCO_3^-)

and carbonate (CO2–3). The ratio of these species depends on factors such as seawater temperature and alkalinity. These different forms of dissolved inorganic carbon are transferred from an ocean's surface to its interior by the ocean's solubility pump.

The resistance of an area of ocean to absorbing atmospheric CO_2 is known as the Revelle factor.

Acidification

Dissolving CO2 in seawater increases the hydrogen ion (H+) concentration in the ocean, and thus decreases ocean pH, as follows:

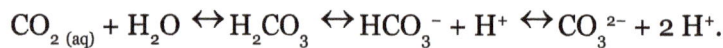

$$CO_{2\ (aq)} + H_2O \leftrightarrow H_2CO_3 \leftrightarrow HCO_3^- + H^+ \leftrightarrow CO_3^{2-} + 2\,H^+.$$

Caldeira and Wickett (2003) placed the rate and magnitude of modern ocean acidification changes in the context of probable historical changes during the last 300 million years.

Since the industrial revolution began, it is estimated that surface ocean pH has dropped by slightly more than 0.1 units on the logarithmic scale of pH, representing about a 29% increase in H+. It is expected to drop by a further 0.3 to 0.5 pH units (an additional doubling to tripling of today's post-industrial acid concentrations) by 2100 as the oceans absorb more anthropogenic CO2, the impacts being most severe for coral reefs and the Southern Ocean. These changes are predicted to continue rapidly as the oceans take up more anthropogenic CO2 from the atmosphere. The degree of change to ocean chemistry, including ocean pH, will depend on the mitigation and emissions pathways society takes.

Although the largest changes are expected in the future, a report from NOAA scientists found large quantities of water undersaturated in aragonite are already upwelling close to the Pacific continental shelf area of North America. Continental shelves play an important role in marine ecosystems since most marine organisms live or are spawned there, and though the study only dealt with the area from Vancouver to Northern California, the authors suggest that other shelf areas may be experiencing similar effects.

Average surface ocean pH				
Time	pH	pH change relative to pre-industrial	Source	H+ concentration change relative to pre-industrial
Pre-industrial (18th century)	8.179		analysed field	
Recent past (1990s)	8.104	−0.075	field	+ 18.9%
Present levels	~8.069	−0.11	field	**+ 28.8%**
2050 (2×CO 2 = 560 ppm)	7.949	−0.230	model	+ 69.8%
2100 (IS92a)	7.824	−0.355	model	+ 126.5%

Rate

One of the first detailed datasets to examine how pH varied over a period of time at a temperate coastal location found that acidification was occurring much faster than previously predicted, with consequences for near-shore benthic ecosystems. Thomas Lovejoy, former chief biodiversity advisor to the World Bank, has suggested that "the acidity of the oceans will more than double in the next 40 years. This rate is 100 times faster than any changes in ocean acidity in the last 20 million years, making it unlikely that marine life can somehow adapt to the changes." It is predicted that, by the year 2100, the level of acidity in the ocean will reach the levels experienced by the earth 20 million years ago.

Current rates of ocean acidification have been compared with the greenhouse event at the Paleocene–Eocene boundary (about 55 million years ago) when surface ocean temperatures rose by 5–6 degrees Celsius. No catastrophe was seen in surface ecosystems, yet bottom-dwelling organisms in the deep ocean experienced a major extinction. The current acidification is on a path to reach levels higher than any seen in the last 65 million years, and the rate of increase is about ten times the rate that preceded the Paleocene–Eocene mass extinction. The current and projected acidification has been described as an almost unprecedented geological event. A National Research Council study released in April 2010 likewise concluded that "the level of acid in the oceans is increasing at an unprecedented rate." A 2012 paper in the journal *Science* examined the geological record in an attempt to find a historical analog for current global conditions as well as those of the future. The researchers determined that the current rate of ocean acidification is faster than at any time in the past 300 million years.

A review by climate scientists at the RealClimate blog, of a 2005 report by the Royal Society of the UK similarly highlighted the centrality of the *rates* of change in the present anthropogenic acidification process, writing:

"The natural pH of the ocean is determined by a need to balance the deposition and burial of CaCO3 on the sea floor against the influx of Ca2+and CO2–3 into the ocean from dissolving rocks on land, called weathering. These processes stabilize the pH of the ocean, by a mechanism called CaCO3 compensation...The point of bringing it up again is to note that if the CO2 concentration of the atmosphere changes more slowly than this, as it always has throughout the Vostok record, the pH of the ocean will be relatively unaffected because CaCO3 compensation can keep up. The [present] fossil fuel acidification is much faster than natural changes, and so the acid spike will be more intense than the earth has seen in at least 800,000 years."

In the 15-year period 1995–2010 alone, acidity has increased 6 percent in the upper 100 meters of the Pacific Ocean from Hawaii to Alaska. According to a statement in July 2012 by Jane Lubchenco, head of the U.S. National Oceanic and Atmospheric Administration "surface waters are changing much more rapidly than initial calculations have suggested. It's yet another reason to be very seriously concerned about the amount of carbon dioxide that is in the atmosphere now and the additional amount we continue to put out."

A 2013 study claimed acidity was increasing at a rate 10 times faster than in any of the evolutionary crises in Earth's history. In a synthesis report published in *Science* in 2015, 22 leading marine scientists stated that CO_2 from burning fossil fuels is changing the oceans' chemistry more rapidly than at any time since the Great Dying, Earth's most severe known extinction event, emphasizing that the 2 °C maximum temperature increase agreed upon by governments reflects too small a cut in emissions to prevent "dramatic impacts" on the world's oceans, with lead author Jean-Pierre Gattuso remarking that "The ocean has been minimally considered at previous climate negotiations. Our study provides compelling arguments for a radical change at the UN conference (in Paris) on climate change".

Calcification

Overview

Changes in ocean chemistry can have extensive direct and indirect effects on organisms and their habitats. One of the most important repercussions of increasing ocean acidity relates to the production of shells and plates out of calcium carbonate (CaCO3). This process is called calcification and is important to the biology and survival of a wide range of marine organisms. Calcification involves the precipitation of dissolved ions into solid CaCO3 structures, such as coccoliths. After they are formed, such structures are vulnerable to dissolution unless the surrounding seawater contains saturating concentrations of carbonate ions (CO_3^{2-}).

Mechanism

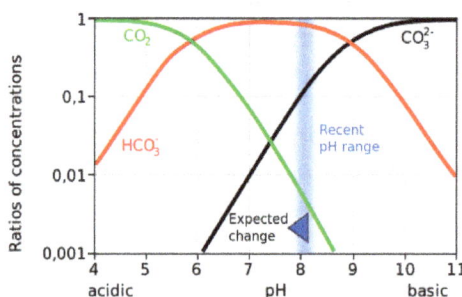

Bjerrum plot: Change in carbonate system of seawater from ocean acidification.

Of the extra carbon dioxide added into the oceans, some remains as dissolved carbon dioxide, while the rest contributes towards making additional bicarbonate (and additional carbonic acid). This also increases the concentration of hydrogen ions, and the percentage increase in hydrogen is larger than the percentage increase in bicarbonate, creating an imbalance in the reaction $HCO_3^- \leftrightarrow CO_3^{2-} + H^+$. To maintain chemical equilibrium, some of the carbonate ions already in the ocean combine with some of the hydrogen ions to make further bicarbonate. Thus the ocean's concentration of carbonate ions is reduced, creating an imbalance in the reaction $Ca^{2+} + CO_3^{2-} \leftrightarrow CaCO_3$, and making the dissolution of formed CaCO3 structures more likely.

These increases in concentrations of dissolved carbon dioxide and bicarbonate, and reduction in carbonate, are shown in a Bjerrum plot.

Saturation State

The saturation state (known as Ω) of seawater for a mineral is a measure of the thermodynamic potential for the mineral to form or to dissolve, and is described by the following equation:

$$\Omega = \frac{\left[Ca^{2+}\right]\left[CO_3^{2-}\right]}{K_{sp}}$$

Here Ω is the product of the concentrations (or activities) of the reacting ions that form the mineral (Ca2+and CO2−3), divided by the product of the concentrations of those ions when the mineral is at equilibrium (Ksp), that is, when the mineral is neither forming nor dissolving. In seawater, a natural horizontal boundary is formed as a result of temperature, pressure, and depth, and is known as the saturation horizon, or lysocline. Above this saturation horizon, Ω has a value greater than 1, and $CaCO_3$ does not readily dissolve. Most calcifying organisms live in such waters. Below this depth, Ω has a value less than 1, and $CaCO_3$ will dissolve. However, if its production rate is high enough to offset dissolution, $CaCO_3$ can still occur where Ω is less than 1. The carbonate compensation depth occurs at the depth in the ocean where production is exceeded by dissolution.

The decrease in the concentration of CO_3^{2-} decreases Ω, and hence makes $CaCO_3$ dissolution more likely.

Calcium carbonate occurs in two common polymorphs (crystalline forms): aragonite and calcite. Aragonite is much more soluble than calcite, so the aragonite saturation horizon is always nearer to the surface than the calcite saturation horizon. This also means that those organisms that produce aragonite may be more vulnerable to changes in ocean acidity than those that produce calcite. Increasing CO_2 levels and the resulting lower pH of seawater decreases the saturation state of $CaCO_3$ and raises the saturation horizons of both forms closer to the surface. This decrease in saturation state is believed to be one of the main factors leading to decreased calcification in marine organisms, as the inorganic precipitation of $CaCO_3$ is directly proportional to its saturation state.

Possible Impacts

Increasing acidity has possibly harmful consequences, such as depressing metabolic rates in jumbo squid, depressing the immune responses of blue mussels, and coral bleaching. However it may benefit some species, for example increasing the growth rate of the sea star, *Pisaster ochraceus*, while shelled plankton species may flourish in altered oceans.

The report "Ocean Acidification Summary for Policymakers 2013" describes research findings and possible impacts.

Impacts on Oceanic Calcifying Organisms

Although the natural absorption of CO_2 by the world's oceans helps mitigate the climatic effects of anthropogenic emissions of CO_2, it is believed that the resulting decrease in pH will have negative consequences, primarily for oceanic calcifying organisms. These span the food chain from autotrophs to heterotrophs and include organisms such as coccolithophores, corals, foraminifera, echinoderms, crustaceans and molluscs. As described above, under normal conditions, calcite and aragonite are stable in surface waters since the carbonate ion is at supersaturating concentrations. However, as ocean pH falls, the concentration of carbonate ions required for saturation to occur increases, and when carbonate becomes undersaturated, structures made of calcium carbonate are vulnerable to dissolution. Therefore, even if there is no change in the rate of calcification, the rate of dissolution of calcareous material increases.

Corals, coccolithophore algae, coralline algae, foraminifera, shellfish and pteropods experience reduced calcification or enhanced dissolution when exposed to elevated CO_2.

The Royal Society published a comprehensive overview of ocean acidification, and its potential consequences, in June 2005. However, some studies have found different response to ocean acidification, with coccolithophore calcification and photosynthesis both increasing under elevated atmospheric pCO_2, an equal decline in primary production and calcification in response to elevated CO_2 or the direction of the response varying between species. A study in 2008 examining a sediment core from the North Atlantic found that while the species composition of coccolithophorids has remained unchanged for the industrial period 1780 to 2004, the calcification of coccoliths has increased by up to 40% during the same time. A 2010 study from Stony Brook University suggested that while some areas are overharvested and other fishing grounds are being restored, because of ocean acidification it may be impossible to bring back many previous shellfish populations. While the full ecological consequences of these changes in calcification are still uncertain, it appears likely that many calcifying species will be adversely affected.

When exposed in experiments to pH reduced by 0.2 to 0.4, larvae of a temperate brittlestar, a relative of the common sea star, fewer than 0.1 percent survived more than eight days. There is also a suggestion that a decline in the coccolithophores may have secondary effects on climate, contributing to global warming by decreasing the Earth's albedo via their effects on oceanic cloud cover. All marine ecosystems on Earth will be exposed to changes in acidification and several other ocean biogeochemical changes.

The fluid in the internal compartments where corals grow their exoskeleton is also extremely important for calcification growth. When the saturation rate of aragonite in the external seawater is at ambient levels, the corals will grow their aragonite crystals rapidly in their internal compartments, hence their exoskeleton grows rapidly. If the level of aragonite in the external seawater is lower than the ambient level, the corals have to work harder to maintain the right balance in the internal compartment. When that happens, the process of growing the crystals slows down, and this slows down the rate of how much their exoskeleton is growing. Depending on how much aragonite is in the surrounding water, the corals may even stop growing because the levels of aragonite are too low to pump in to the internal compartment. They could even dissolve faster than they can make the crystals to their skeleton, depending on the aragonite levels in the surrounding water.

Ocean acidification may force some organisms to reallocate resources away from productive endpoints such as growth in order to maintain calcification.

In some places carbon dioxide bubbles out from the sea floor, locally changing the pH and other aspects of the chemistry of the seawater. Studies of these carbon dioxide seeps have documented a variety of responses by different organisms. Coral reef communities located near carbon dioxide seeps are of particular interest because of the sensitivity of some corals species to acidification. In Papua New Guinea, declining pH caused by carbon dioxide seeps is associated with declines in coral species diversity. However, in Palau carbon dioxide seeps are not associated with reduced species diversity of corals, although bioerosion of coral skeletons is much higher at low pH sites.

Other Biological Impacts

Aside from the slowing and/or reversing of calcification, organisms may suffer other adverse effects, either indirectly through negative impacts on food resources, or directly as reproductive or physiological effects. For example, the elevated oceanic levels of CO_2 may produce CO2-induced acidification of body fluids, known as hypercapnia. Also, increasing ocean acidity is believed to have a range of direct consequences. For example, increasing acidity has been observed to: reduce metabolic rates in jumbo squid; depress the immune responses of blue mussels; and make it harder for juvenile clownfish to tell apart the smells of non-predators and predators, or hear the sounds of their predators. This is possibly because ocean acidification may alter the acoustic properties of seawater, allowing sound to propagate further, and increasing ocean noise. Calcium carbonate ions are very important when it comes to building organisms as they use calcium to build skeletons and shells. OA affects ocean species to a varying degree. Many plants do well with high CO_2 levels, but organisms that calcify like clams, mussels, corals and sea urchins might not do so well because everything in the ocean is connected including the food web. Since OA is such a large event because the ocean takes up 71% of the earth's surface OA is being referred to as climate change because it will affect the whole planet if the ocean becomes acidic. This impacts all animals that use sound for echolocation or communication. Atlantic longfin squid eggs took longer to hatch in acidified water, and the squid's statolith was smaller and malformed in animals placed in sea water with a lower pH. The lower PH was simulated with 20-30 times the normal amount of CO_2. However, as with calcification, as yet there is not a full understanding of these processes in marine organisms or ecosystems.

Another possible effect would be an increase in red tide events, which could contribute to the accumulation of toxins (domoic acid, brevetoxin, saxitoxin) in small organisms such as anchovies and shellfish, in turn increasing occurrences of amnesic shellfish poisoning, neurotoxic shellfish poisoning and paralytic shellfish poisoning.

Nonbiological Impacts

Leaving aside direct biological effects, it is expected that ocean acidification in the future will lead to a significant decrease in the burial of carbonate sediments for several centuries, and even the dissolution of existing carbonate sediments. This will cause an elevation of ocean alkalinity, leading to the enhancement of the ocean as a reservoir for CO_2 with implications for climate change as more CO_2 leaves the atmosphere for the ocean.

Impact on Human Industry

The threat of acidification includes a decline in commercial fisheries and in the Arctic tourism industry and economy. Commercial fisheries are threatened because acidification harms calcifying organisms which form the base of the Arctic food webs.

Pteropods and brittle stars both form the base of the Arctic food webs and are both seriously damaged from acidification. Pteropods shells dissolve with increasing acidification and the brittle stars lose muscle mass when re-growing appendages. For pteropods to create shells they require aragonite which is produced through carbonate ions and dissolved calcium. Pteropods are severely affected because increasing acidification levels have steadily decreased the amount of water supersaturated with carbonate which is needed for aragonite creation. Arctic waters are changing so rapidly that they will become undersaturated with aragonite as early as 2016. Additionally the brittle star's eggs die within a few days when exposed to expected conditions resulting from Arctic acidification. Acidification threatens to destroy Arctic food webs from the base up. Arctic food webs are considered simple, meaning there are few steps in the food chain from small organisms to larger predators. For example, pteropods are "a key prey item of a number of higher predators - larger plankton, fish, seabirds, whales" Both pteropods and sea stars serve as a substantial food source and their removal from the simple food web would pose a serious threat to the whole ecosystem. The effects on the calcifying organisms at the base of the food webs could potentially destroy fisheries. The value of fish caught from US commercial fisheries in 2007 was valued at $3.8 billion and of that 73% was derived from calcifiers and their direct predators. Other organisms are directly harmed as a result of acidification. For example, decrease in the growth of marine calcifiers such as the American Lobster, Ocean Quahog, and scallops means there is less shellfish meat available for sale and consumption. Red king crab fisheries are also at a serious threat because crabs are calcifiers and rely on carbonate ions for shell development. Baby red king crab when exposed to increased acidification levels experienced 100% mortality after 95 days. In 2006 Red King Cab accounted for 23% of the total guideline harvest levels and a serious decline in red crab population would threaten the crab harvesting industry. Several ocean goods and services are likely to be undermined by future ocean acidification potentially affecting the livelihoods of some 400 to 800 million people depending upon the emission scenario.

Impact on Indigenous Peoples

Acidification could damage the Arctic tourism economy and affect the way of life of indigenous peoples. A major pillar of Arctic tourism is the sport fishing and hunting industry. The sport fishing industry is threatened by collapsing food webs which provide food for the prized fish. A decline in tourism lowers revenue input in the area, and threatens the economies that are increasingly dependent on tourism. Acidification is not merely a threat but has significantly declined whole fish populations. For example, In Scandinavia studies conducted on acidic water revealed that 15% of species populations had disappeared and that many more populations were limited in numbers or declining. The rapid decrease or disappearance of marine life could also affect the diet of Indigenous peoples.

Possible Responses

Reducing CO$_2$ Emissions

Members of the InterAcademy Panel recommended that by 2050, global anthropogenic CO$_2$ emissions be reduced less than 50% of the 1990 level. The 2009 statement also called on world leaders to:

- Acknowledge that ocean acidification is a direct and real consequence of increasing atmospheric CO$_2$ concentrations, is already having an effect at current concentrations, and is likely to cause grave harm to important marine ecosystems as CO$_2$ concentrations reach 450 [parts-per-million (ppm)] and above;

- [...] Recognise that reducing the build up of CO$_2$ in the atmosphere is the only practicable solution to mitigating ocean acidification;

- [...] Reinvigorate action to reduce stressors, such as overfishing and pollution, on marine ecosystems to increase resilience to ocean acidification.

Stabilizing atmospheric CO$_2$ concentrations at 450 ppm would require near-term emissions reductions, with steeper reductions over time.

The German Advisory Council on Global Change stated:

In order to prevent disruption of the calcification of marine organisms and the resultant risk of fundamentally altering marine food webs, the following guard rail should be obeyed: the pH of near surface waters should not drop more than 0.2 units below the pre-industrial average value in any larger ocean region (nor in the global mean).

One policy target related to ocean acidity is the magnitude of future global warming. Parties to the United Nations Framework Convention on Climate Change (UNFCCC) adopted a target of limiting warming to below 2 °C, relative to the pre-industrial level. Meeting this target would require substantial reductions in anthropogenic CO_2 emissions.

Limiting global warming to below 2 °C would imply a reduction in surface ocean pH of 0.16 from pre-industrial levels. This would represent a substantial decline in surface ocean pH.

Climate Engineering

Climate engineering (mitigating temperature or pH effects of emissions) has been proposed as a possible response to ocean acidification. The IAP (2009) statement cautioned against climate engineering as a policy response:

Mitigation approaches such as adding chemicals to counter the effects of acidification are likely to be expensive, only partly effective and only at a very local scale, and may pose additional unanticipated risks to the marine environment. There has been very little research on the feasibility and impacts of these approaches. Substantial research is needed before these techniques could be applied.

Reports by the WGBU (2006), the UK's Royal Society (2009), and the US National Research Council (2011) warned of the potential risks and difficulties associated with climate engineering.

Iron Fertilization

Iron fertilization of the ocean could stimulate photosynthesis in phytoplankton. The phytoplankton would convert the ocean's dissolved carbon dioxide into carbohydrate and oxygen gas, some of which would sink into the deeper ocean before oxidizing. More than a dozen open-sea experiments confirmed that adding iron to the ocean increases photosynthesis in phytoplankton by up to 30 times. While this approach has been proposed as a potential solution to the ocean acidification problem, mitigation of surface ocean acidification might increase acidification in the less-inhabited deep ocean.

A report by the UK's Royal Society (2009) reviewed the approach for effectiveness, affordability, timeliness and safety. The rating for affordability was "medium", or "not expected to be very cost-effective." For the other three criteria, the ratings ranged from "low" to "very low" (i.e., not good). For example, in regards to safety, the report found a "[high] potential for undesirable ecological side effects," and that ocean fertilization "may increase anoxic regions of ocean ('dead zones')."

Carbon Negative Fuels

Carbonic acid can be extracted from seawater as carbon dioxide for use in making synthetic fuel. If the resulting flue exhaust gas was subject to carbon capture, then the process would be carbon negative over time, resulting in permanent extraction of inorganic carbon from seawater and the atmosphere with which seawater is in equilibrium. Based on the energy requirements, this process was estimated to cost about $50 per tonne of CO_2.

Megadrought

A megadrought (or mega-drought) is a prolonged drought lasting two decades or longer. Past megadroughts have been associated with persistent multiyear La Niña conditions (cooler than normal water temperatures in the tropical eastern Pacific Ocean).

The term megadrought is generally used to describe the length of a drought, and not its acute intensity. In scientific literature the term is used to describe decades-long droughts or multi-decadal droughts. Multiyear droughts of less than a decade, such as the Dust Bowl drought of the 1930s, are generally not described as megadroughts even though they are of a long duration. In popular literature multiyear or even single year droughts are occasionally described as megadroughts based upon their severity, the economic damage they inflict or other criteria, but this is the exception and not the rule.

Impact

Megadroughts have historically led to the mass migration of humans away from drought affected lands, resulting in a significant population decline from pre-drought levels. They are suspected of playing a primary role in the collapse of several pre-industrial civilizations, including the Anasazi of the North American Southwest, the Khmer Empire of Cambodia, the Mayan of Mesoamerica, the Tiwanaku of Bolivia, and the Yuan Dynasty of China.

The African Sahel region in particular has suffered multiple megadroughts throughout history, with the most recent lasting from approximately 1400 AD to 1750 AD. North America experienced at least four megadroughts during the Medieval Warm Period.

Historical Evidence

Montezuma Bald Cypress tree, 900 years old

There are several sources for establishing the past occurrence and frequency of megadroughts, including:

- When megadroughts occur, lakes dry up and trees and other plants grow in the dry lake beds. When the drought ends the lakes refill, when this happens the trees are submerged and die. In some locations these trees have remained preserved and can be studied giving accurate radio-carbon dates, and the tree rings of the same long dead trees can be studied. Such trees have been found in Mono and Tenaya lakes in California, Lake Bosumtwi in Ghana; and various other lakes.

- Dendrochronology, the dating and study of annual rings in trees. The tree-ring data indicate that the Western states have experienced droughts that lasted ten times longer than anything the modern U.S. has seen. Based on annual tree rings, NOAA has recorded patterns of drought covering most of the U.S. for every year since 1700. Certain species of trees have given evidence over a longer period, in particular Montezuma Cypress and Bristlecone pine trees. The University of Arkansas has produced a 1238-year tree-ring based chronology of weather condition in central Mexico by examining core samples taken from living Montezuma Cypress trees.

- Sediment core samples taken at the volcanic caldera in Valles Caldera, New Mexico and other locations. The cores from Valles Caldera go back 550,000 years and show evidence of megadroughts that lasted as long as 1000 years during the mid-Pleistocene Epoch during which summer rains were almost non-existent. Plant and pollen remains found in core samples from the bottom of lakes have been also studied and added to the record.

- Fossil corals on Palmyra Atoll. Using the relationship between tropical Pacific sea surface temperatures and the oxygen isotope ratio in living corals to convert fossil coral records into sea surface temperatures. This has been used to establish the occurrence and frequency of La Niña conditions.

- During a 200-year mega drought in the Sierra Nevada that lasted from the 9th and 12th centuries, trees would grow on newly exposed shoreline at Fallen Leaf Lake, then as the lake grew once again, the trees were preserved under cold water.

Acid Rain

Acid rain is a rain or any other form of precipitation that is unusually acidic, meaning that it possesses elevated levels of hydrogen ions (low pH). It can have harmful effects on plants, aquatic animals and infrastructure. Acid rain is caused by emissions of sulfur dioxide and nitrogen oxide, which react with the water molecules in the atmosphere to produce acids. Some Governments have made efforts since the 1970s to reduce the release of sulfur dioxide and nitrogen oxide into the atmosphere with positive results. Nitrogen oxides can also be produced naturally by lightning strikes, and sulfur dioxide is produced by volcanic eruptions. The chemicals in acid rain can cause paint to peel, corrosion of steel structures such as bridges, and weathering of stone buildings and statues.

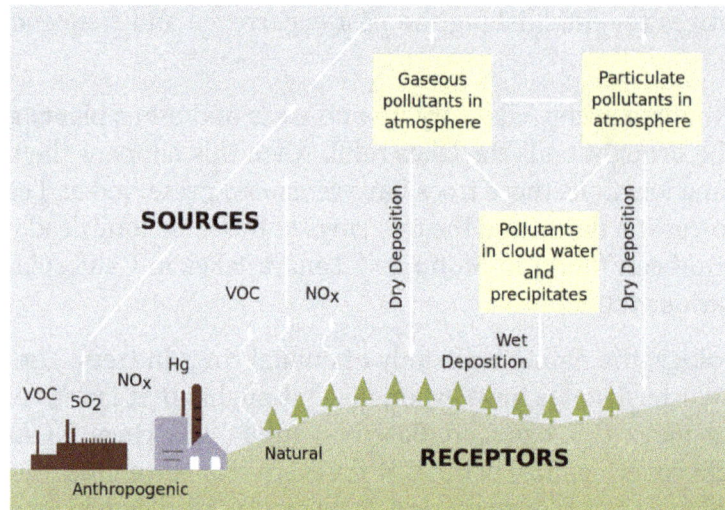

Processes involved in acid deposition (note that only SO_2 and NO_x play a significant role in acid rain).

Acid clouds can grow on SO_2 emissions from refineries, as seen here in Curaçao.

Definition

"Acid rain" is a popular term referring to the deposition of a mixture from wet (rain, snow, sleet, fog, cloudwater, and dew) and dry (acidifying particles and gases) acidic components. Distilled water, once carbon dioxide is removed, has a neutral pH of 7. Liquids with a pH less than 7 are acidic, and those with a pH greater than 7 are alkaline. "Clean" or unpolluted rain has an acidic pH, but usually no lower than 5.7, because carbon dioxide and water in the air react together to form carbonic acid, a weak acid according to the following reaction:

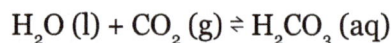

$$H_2O \text{ (l)} + CO_2 \text{ (g)} \rightleftharpoons H_2CO_3 \text{ (aq)}$$

Carbonic acid then can ionize in water forming low concentrations of hydronium and carbonate ions:

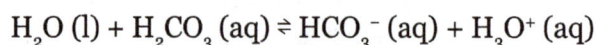

$$H_2O \text{ (l)} + H_2CO_3 \text{ (aq)} \rightleftharpoons HCO_3^- \text{ (aq)} + H_3O^+ \text{ (aq)}$$

However, unpolluted rain can also contain other chemicals which affect its pH (acidity level). A common example is nitric acid produced by electric discharge in the atmosphere such as lightning. Acid deposition as an environmental issue would include additional acids to H_2CO_3.

History

The corrosive effect of polluted, acidic city air on limestone and marble was noted in the 17th century by John Evelyn, who remarked upon the poor condition of the Arundel marbles. Since the Industrial Revolution, emissions of sulfur dioxide and nitrogen oxides into the atmosphere have increased. In 1852, Robert Angus Smith was the first to show the relationship between acid rain and atmospheric pollution in Manchester, England.

Though acidic rain was discovered in 1853, it was not until the late 1960s that scientists began widely observing and studying the phenomenon. The term "acid rain" was coined in 1872 by Robert Angus Smith. Canadian Harold Harvey was among the first to research a "dead" lake. Public awareness of acid rain in the U.S increased in the 1970s after The New York Times published reports from the Hubbard Brook Experimental Forest in New Hampshire of the myriad deleterious environmental effects shown to result from it.

Occasional pH readings in rain and fog water of well below 2.4 have been reported in industrialized areas. Industrial acid rain is a substantial problem in China and Russia and areas downwind from them. These areas all burn sulfur-containing coal to generate heat and electricity.

The problem of acid rain has not only increased with population and industrial growth, but has become more widespread. The use of tall smokestacks to reduce local pollution has contributed to the spread of acid rain by releasing gases into regional atmospheric circulation. Often deposition occurs a considerable distance downwind of the emissions, with mountainous regions tending to receive the greatest deposition (simply because of their higher rainfall). An example of this effect is the low pH of rain which falls in Scandinavia.

History of Acid Rain in the United States

Since 1998, Harvard University wraps some of the bronze and marble statues on its campus, such as this "Chinese stele", with waterproof covers every winter, in order to protect them from erosion caused by acid rain and acid snow

In 1980, the U.S. Congress passed an Acid Deposition Act. This Act established an 18-year assessment and research program under the direction of the National Acidic Precipitation Assessment Program (NAPAP). NAPAP looked at the entire problem from a scientific perspective. It enlarged a network of monitoring sites to determine how acidic the precipitation actually was, and to determine long-term trends, and established a network for dry deposition. It looked at the effects of acid rain and funded research on the effects of acid precipitation on freshwater and terrestrial ecosystems, historical buildings, monuments, and building materials. It also funded extensive studies on atmospheric processes and potential control programs.

From the start, policy advocates from all sides attempted to influence NAPAP activities to support their particular policy advocacy efforts, or to disparage those of their opponents. For the U.S. Government's scientific enterprise, a significant impact of NAPAP were lessons learned in the assessment process and in environmental research management to a relatively large group of scientists, program managers and the public.

In 1991, DENR provided its first assessment of acid rain in the United States. It reported that 5% of New England Lakes were acidic, with sulfates being the most common problem. They noted that 2% of the lakes could no longer support Brook Trout, and 6% of the lakes were unsuitable for the survival of many species of minnow. Subsequent Reports to Congress have documented chemical changes in soil and freshwater ecosystems, nitrogen saturation, decreases in amounts of nutrients in soil, episodic acidification, regional haze, and damage to historical monuments.

Meanwhile, in 1989, the U.S. Congress passed a series of amendments to the Clean Air Act. Title IV of these amendments established the Acid Rain Program, a cap and trade system designed to control emissions of sulfur dioxide and nitrogen oxides. Title IV called for a total reduction of about 10 million tons of SO_2 emissions from power plants. It was implemented in two phases. Phase I began in 1995, and limited sulfur dioxide emissions from 110 of the largest power plants to a combined total of 8.7 million tons of sulfur dioxide. One power plant in New England (Merrimack) was in Phase I. Four other plants (Newington, Mount Tom, Brayton Point, and Salem Harbor) were added under other provisions of the program. Phase II began in 2000, and affects most of the power plants in the country.

During the 1990s, research continued. On March 10, 2005, EPA issued the Clean Air Interstate Rule (CAIR). This rule provides states with a solution to the problem of power plant pollution that drifts from one state to another. CAIR will permanently cap emissions of SO_2 and NO_x in the eastern United States. When fully implemented, CAIR will reduce SO_2 emissions in 28 eastern states and the District of Columbia by over 70% and NO_x emissions by over 60% from 2003 levels.

Overall, the program's cap and trade program has been successful in achieving its goals. Since the 1990s, SO_2 emissions have dropped 40%, and according to the Pacific Research Institute, acid rain levels have dropped 65% since 1976. Conventional regulation was used in the European Union, which saw a decrease of over 70% in SO_2 emissions during the same time period.

In 2007, total SO_2 emissions were 8.9 million tons, achieving the program's long-term goal ahead of the 2010 statutory deadline.

In 2007 the EPA estimated that by 2010, the overall costs of complying with the program for businesses and consumers would be $1 billion to $2 billion a year, only one fourth of what was originally predicted. Forbes says: *In 2010, by which time the cap and trade system had been augmented by the George W. Bush administration's Clean Air Interstate Rule, SO2 emissions had fallen to 5.1 million tons.*

Emissions of Chemicals Leading to Acidification

The most important gas which leads to acidification is sulfur dioxide. Emissions of nitrogen oxides which are oxidized to form nitric acid are of increasing importance due to stricter controls on emissions of sulfur containing compounds. 70 Tg(S) per year in the form of SO_2 comes from fossil fuel combustion and industry, 2.8 Tg(S) from wildfires and 7–8 Tg(S) per year from volcanoes.

Natural Phenomena

The principal natural phenomena that contribute acid-producing gases to the atmosphere are emissions from volcanoes. Thus, for example, fumaroles from the Laguna Caliente crater of Poás Volcano create extremely high amounts of acid rain and fog, with acidity as high as a pH of 2, clearing an area of any vegetation and frequently causing irritation to the eyes and lungs of inhabitants in nearby settlements. Acid-producing gasses are also created by biological processes that occur on the land, in wetlands, and in the oceans. The major biological source of sulfur containing compounds is dimethyl sulfide.

Nitric acid in rainwater is an important source of fixed nitrogen for plant life, and is also produced by electrical activity in the atmosphere such as lightning.

Acidic deposits have been detected in glacial ice thousands of years old in remote parts of the globe.

Soils of coniferous forests are naturally very acidic due to the shedding of needles, and the results of this phenomenon should not be confused with acid rain.

Human Activity

The coal-fired Gavin Power Plant in Cheshire, Ohio

The principal cause of acid rain is sulfur and nitrogen compounds from human sources, such as electricity generation, factories, and motor vehicles. Electrical power generation using coal is among the greatest contributors to gaseous pollutions that are responsible for acidic rain. The gases can be carried hundreds of kilometers in the atmosphere before they are converted to acids and deposited. In the past, factories had short funnels to let out smoke but this caused many problems locally; thus, factories now have taller smoke funnels. However, dispersal from these taller stacks causes pollutants to be carried farther, causing widespread ecological damage.

Chemical Processes

Combustion of fuels produces sulfur dioxide and nitric oxides. They are converted into sulfuric acid and nitric acid.

Gas Phase Chemistry

In the gas phase sulfur dioxide is oxidized by reaction with the hydroxyl radical via an intermolecular reaction:

$$SO_2 + OH\cdot \rightarrow HOSO_2\cdot$$

which is followed by:

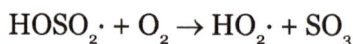

$$HOSO_2\cdot + O_2 \rightarrow HO_2\cdot + SO_3$$

In the presence of water, sulfur trioxide (SO_3) is converted rapidly to sulfuric acid:

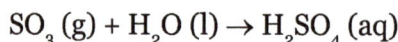

$$SO_3\ (g) + H_2O\ (l) \rightarrow H_2SO_4\ (aq)$$

Nitrogen dioxide reacts with OH to form nitric acid:

$$NO_2 + OH\cdot \rightarrow HNO_3$$

Chemistry in Cloud Droplets

When clouds are present, the loss rate of SO_2 is faster than can be explained by gas phase chemistry alone. This is due to reactions in the liquid water droplets.

Hydrolysis

Sulfur dioxide dissolves in water and then, like carbon dioxide, hydrolyses in a series of equilibrium reactions:

$$SO_2\ (g) + H_2O \rightleftharpoons SO_2\cdot H_2O$$

$$SO_2\cdot H_2O \rightleftharpoons H^+ + HSO_3^-$$

$$HSO_3^- \rightleftharpoons H^+ + SO_3^{2-}$$

Oxidation

There are a large number of aqueous reactions that oxidize sulfur from S(IV) to S(VI), leading to the formation of sulfuric acid. The most important oxidation reactions are with ozone, hydrogen peroxide and oxygen (reactions with oxygen are catalyzed by iron and manganese in the cloud droplets).

Acid Deposition

Wet Deposition

Wet deposition of acids occurs when any form of precipitation (rain, snow, and so on.) removes acids from the atmosphere and delivers it to the Earth's surface. This can result from the deposition of acids produced in the raindrops or by the precipitation removing the acids either in clouds or below clouds. Wet removal of both gases and aerosols are both of importance for wet deposition.

Dry Deposition

Acid deposition also occurs via dry deposition in the absence of precipitation. This can be responsible for as much as 20 to 60% of total acid deposition. This occurs when particles and gases stick to the ground, plants or other surfaces.

Adverse Effects

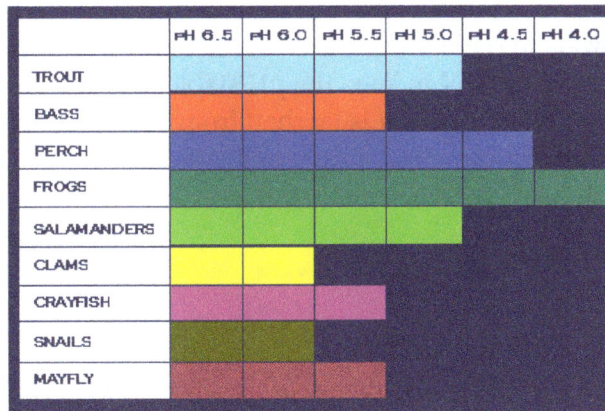

	pH 6.5	pH 6.0	pH 5.5	pH 5.0	pH 4.5	pH 4.0
TROUT						
BASS						
PERCH						
FROGS						
SALAMANDERS						
CLAMS						
CRAYFISH						
SNAILS						
MAYFLY						

This chart shows that not all fish, shellfish, or the insects that they eat can tolerate the same amount of acid; for example, frogs can tolerate water that is more acidic (i.e., has a lower pH) than trout.

Acid rain has been shown to have adverse impacts on forests, freshwaters and soils, killing insect and aquatic life-forms as well as causing damage to buildings and having impacts on human health.

Surface Waters and Aquatic Animals

Both the lower pH and higher aluminium concentrations in surface water that occur as a result of acid rain can cause damage to fish and other aquatic animals. At pHs lower than 5 most fish eggs will not hatch and lower pHs can kill adult fish. As lakes and rivers become more acidic biodiversity is reduced. Acid rain has eliminated insect life and some fish species, including the brook trout in some lakes, streams, and creeks in geographically sensitive areas, such as the Adirondack Mountains of the United States. However, the extent to which acid rain contributes directly or indirectly via runoff from the catchment to lake and river acidity (i.e., depending on characteristics of the surrounding watershed) is variable. The United States Environmental Protection Agency's (EPA) website states: "Of the lakes and streams surveyed, acid rain caused acidity in 75% of the acidic lakes and about 50% of the acidic streams".

Soils

Soil biology and chemistry can be seriously damaged by acid rain. Some microbes are unable to tolerate changes to low pH and are killed. The enzymes of these microbes are denatured (changed in shape so they no longer function) by the acid. The hydronium ions of acid rain also mobilize toxins such as aluminium, and leach away essential nutrients and minerals such as magnesium.

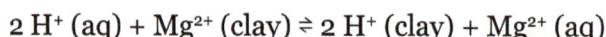

$$2 H^+ (aq) + Mg^{2+} (clay) \rightleftharpoons 2 H^+ (clay) + Mg^{2+} (aq)$$

Soil chemistry can be dramatically changed when base cations, such as calcium and magnesium, are leached by acid rain thereby affecting sensitive species, such as sugar maple (Acer saccharum).

Forests and Other Vegetation

Adverse effects may be indirectly related to acid rain, like the acid's effects on soil or high concentration of gaseous precursors to acid rain. High altitude forests are especially vulnerable as they are often surrounded by clouds and fog which are more acidic than rain.

Other plants can also be damaged by acid rain, but the effect on food crops is minimized by the application of lime and fertilizers to replace lost nutrients. In cultivated areas, limestone may also be added to increase the ability of the soil to keep the pH stable, but this tactic is largely unusable in the case of wilderness lands. When calcium is leached from the needles of red spruce, these trees become less cold tolerant and exhibit winter injury and even death.

Ocean Acidification

Human Health Effects

Acid rain does not directly affect human health. The acid in the rainwater is too dilute to have direct adverse effects. However, the particulates responsible for acid rain (sulfur dioxide and nitrogen oxides) do have an adverse effect. Increased amounts of fine particulate matter in the air do contribute to heart and lung problems including asthma and bronchitis.

Other Adverse Effects

Effect of acid rain on statues

Acid rain and weathering

Acid rain can damage buildings, historic monuments, and statues, especially those made of rocks, such as limestone and marble, that contain large amounts of calcium carbonate. Acids in the rain react with the calcium compounds in the stones to create gypsum, which then flakes off.

$$CaCO_3 (s) + H_2SO_4 (aq) \rightleftharpoons CaSO_4 (s) + CO_2 (g) + H_2O (l)$$

The effects of this are commonly seen on old gravestones, where acid rain can cause the inscriptions to become completely illegible. Acid rain also increases the corrosion rate of metals, in particular iron, steel, copper and bronze.

Affected Areas

Places significantly impacted by acid rain around the globe include most of eastern Europe from Poland northward into Scandinavia, the eastern third of the United States, and southeastern Canada. Other affected areas include the southeastern coast of China and Taiwan.

Prevention Methods

Technical Solutions

Many coal-firing power stations use flue-gas desulfurization (FGD) to remove sulfur-containing gases from their stack gases. For a typical coal-fired power station, FGD will remove 95% or more of the SO_2 in the flue gases. An example of FGD is the wet scrubber which is commonly used. A wet scrubber is basically a reaction tower equipped with a fan that extracts hot smoke stack gases from a power plant into the tower. Lime or limestone in slurry form is also injected into the tower to mix with the stack gases and combine with the sulfur dioxide present. The calcium carbonate of the limestone produces pH-neutral calcium sulfate that is physically removed from the scrubber. That is, the scrubber turns sulfur pollution into industrial sulfates.

In some areas the sulfates are sold to chemical companies as gypsum when the purity of calcium sulfate is high. In others, they are placed in landfill. However, the effects of acid rain can last for generations, as the effects of pH level change can stimulate the continued leaching of undesirable chemicals into otherwise pristine water sources, killing off vulnerable insect and fish species and blocking efforts to restore native life.

Fluidized bed combustion also reduces the amount of sulfur emitted by power production.

Vehicle emissions control reduces emissions of nitrogen oxides from motor vehicles.

International Treaties

A number of international treaties on the long-range transport of atmospheric pollutants have been agreed for example, the 1985 Helsinki Protocol on the Reduction of Sulphur Emissions under the Convention on Long-Range Transboundary Air Pollution. Canada and the US signed the Air Quality Agreement in 1991. Most European countries and Canada have signed the treaties.

Emissions Trading

In this regulatory scheme, every current polluting facility is given or may purchase on an open market an emissions allowance for each unit of a designated pollutant it emits. Operators can then install pollution control equipment, and sell portions of their emissions allowances they no longer need for their own operations, thereby recovering some of the capital cost of their investment in such equipment. The intention is to give operators economic incentives to install pollution controls.

The first emissions trading market was established in the United States by enactment of the Clean Air Act Amendments of 1990. The overall goal of the Acid Rain Program established by the Act is to achieve significant environmental and public health benefits through reductions in emissions of sulfur dioxide (SO_2) and nitrogen oxides (NO_x), the primary causes of acid rain. To achieve this goal at the lowest cost to society, the program employs both regulatory and market based approaches for controlling air pollution.

Rain Dust

Rain dust or snow dust, traditionally known as muddy rain, red rain, or coloured rain, is a variety of rain (or any other form of precipitation) which contains enough desert dust for the dust to be visible without using a microscope.

The leaves of this Schefflera show the dust stains of a rain dust (near Paris, France)

Geography

Rain dust is common in the Western Mediterranean, where the dust supply comes from the atmospheric depressions going through the northern part of North Africa. The main sources of desert dust reach the Iberian Peninsula and the Balearic Islands in the form of dust transported by wind or rain from the Western Sahara, Atlas Mountains in Morocco and Central Algeria.

Mud rains are relatively frequent and had been increasing in early 1990s in the Mediterranean Basin.

It also occurs in arid desert regions of North America such as west Texas or Arizona. It occasionally happens in the grasslands as it did in Bexar County, Texas on March 18, 2008.

Dust Composition

The rain dust is very alkaline. Some of the large particles contain mixtures of chemicals such as sulfate and sea salt (chiefly with sodium, chlorine and magnesium). Major minerals in order of decreasing abundance are: illite, quartz, smectite, palygorskite, kaolinite, calcite, dolomite and feldspars. In Majorca a study finds that the size, by volume, 89% of the particles from rain dust fraction corresponded to silt (between 0.002 mm and 0.063 mm) and that there was virtually no clay sized particles (less than 0.29%).

Importance

- The particulates that rain dust carries are important for the formation of long-term soil, counteracting, in large part the effects of soil erosion. The amount of solids in rain dust have been estimated at 5.3 grams per m² (in a study made in Montseny, Catalonia) in this location the dust provides 34% of the calcium needed by the holm oak. The amount of the deposition of dust particles is highly variable depending on the year.

- Saharan dust significantly increases the pH of rain water. This may counteract the effects of acid rain.

- Radioactivity from the Chernobyl Disaster was carried by rain dust to Greece in 2000.

Blood/Red Rain

Rain dust is the most common cause of blood rain.

Red rain is however not always a rain dust.

Smog

Smog in New York City as viewed from the World Trade Center in 1978

German road sign until 2008, *Verkehrsverbot bei Smog* (No traffic allowed in smog conditions)

Smog is a type of air pollutant. The word "smog" was coined in the early 20th century as a portmanteau of the words smoke and fog to refer to smoky fog. The word was then intended to refer to what was sometimes known as pea soup fog, a familiar and serious problem in London from the 19th century to the mid 20th century. This kind of visible air pollution is composed of nitrogen oxides, sulfur oxides, ozone, smoke or particulates among others (less visible pollutants include carbon monoxide, CFCs and radioactive sources). Man-made smog is derived from coal emissions, vehicular emissions, industrial emissions, forest and agricultural fires and photochemical reactions of these emissions.

Modern smog, as found for example in Los Angeles, is a type of air pollution derived from vehicular emission from internal combustion engines and industrial fumes that react in the atmosphere with sunlight to form secondary pollutants that also combine with the primary emissions to form photochemical smog. In certain other cities, such as Delhi, smog severity is often aggravated by stubble burning in neighboring agricultural areas. The atmospheric pollution levels of Los Angeles, Beijing, Delhi, Mexico City, Tehran and other cities are increased by inversion that traps pollution close to the ground. It is usually highly toxic to humans and can cause severe sickness, shortened life or death.

Etymology

Coinage of the term "smog" is generally attributed to Dr. Henry Antoine Des Voeux in his 1905 paper, "Fog and Smoke" for a meeting of the Public Health Congress. The July 26, 1905 edition of the London newspaper *Daily Graphic* quoted Des Voeux, "He said it required no science to see that there was something produced in great cities which was not found in the country, and that was smoky fog, or what was known as 'smog.'" The following day the newspaper stated that "Dr. Des Voeux did a public service in coining a new word for the London fog." However, this is predated by a *Los Angeles Times* article of January 19, 1893, in which the word is attributed to "a witty English writer."

Causes

Coal

Coal fires, used to heat individual buildings or in a power-producing plant, can emit significant clouds of smoke that contributes to smog. Air pollution from this source has been reported in England since the Middle Ages. London, in particular, was notorious up through the mid-20th century for its coal-caused smogs, which were nicknamed 'pea-soupers.' Air pollution of this type is still a problem in areas that generate significant smoke from burning coal, as witnessed by the 2013 autumnal smog in Harbin, China, which closed roads, schools, and the airport.

Transportation Emissions

Traffic emissions – such as from trucks, buses, and automobiles – also contribute. Airborne byproducts from vehicle exhaust systems cause air pollution and are a major ingredient in the creation of smog in some large cities.

The major culprits from transportation sources are carbon monoxide (CO), nitrogen oxides (NO and NO_x), volatile organic compounds, sulfur dioxide, and hydrocarbons. (Hydrocarbons are the main components of petroleum fuels such as gasoline and diesel fuel.) These molecules react with sunlight, heat, ammonia, moisture, and other compounds to form the noxious vapors, ground level ozone, and particles that comprise smog.

Photochemical Smog

Photochemical smog is the chemical reaction of sunlight, nitrogen oxides and volatile organic compounds in the atmosphere, which leaves airborne particles and ground-level ozone. This noxious mixture of air pollutants may include the following:

- Aldehydes

- Nitrogen oxides, particularly nitric oxide and nitrogen dioxide

- Peroxyacyl nitrates

- Tropospheric ozone

- Volatile organic compounds

A primary pollutant is an air pollutant emitted directly from a source. A secondary pollutant is not directly emitted as such, but forms when other pollutants (primary pollutants) react in the atmosphere. Examples of a secondary pollutant include ozone, which is formed when hydrocarbons (HC) and nitrogen oxides (NO_x) combine in the presence of sunlight; nitrogen dioxide (NO_2), which is formed as nitric oxide (NO) combines with oxygen in the air; and acid rain, which is formed when sulfur dioxide or nitrogen oxides react with water. All of these harsh chemicals are usually highly reactive and oxidizing. Photochemical smog is therefore considered to be a problem of modern industrialization. It is present in all modern cities, but it is more common in cities with sunny, warm, dry climates and a large number of motor vehicles. Because it travels with the wind, it can affect sparsely populated areas as well.

The link between automotive exhaust and photochemical smog was discovered in the 1950s by Arie Haagen-Smit.

Characteristic coloration for smog in California in the beige cloud bank behind the Golden Gate Bridge. The brown coloration is due to the NO_x in the photochemical smog.

Natural Causes

An erupting volcano can also emit high levels of sulphur dioxide along with a large quantity of particulate matter; two key components to the creation of smog. However, the smog created as a result of a volcanic eruption is often known as vog to distinguish it as a natural occurrence.

The radiocarbon content of some plant life has been linked to the distribution of smog in some areas. For example, the creosote bush in the Los Angeles area has been shown to have an effect on smog distribution that is more than fossil fuel combustion alone.

Health Effects

Highland Park Optimist Club wearing smog-gas masks at banquet, Los Angeles, circa 1954

Smog is a serious problem in many cities and continues to harm human health. Ground-level ozone, sulfur dioxide, nitrogen dioxide and carbon monoxide are especially harmful for senior citizens, children, and people with heart and lung conditions such as emphysema, bronchitis, and asthma. It can inflame breathing passages, decrease the lungs' working capacity, cause shortness of breath, pain when inhaling deeply, wheezing, and coughing. It can cause eye and nose irritation and it dries out the protective membranes of the nose and throat and interferes with the body's ability to fight infection, increasing susceptibility to illness. Hospital admissions and respiratory deaths often increase during periods when ozone levels are high.

Levels of Unhealthy Exposure

The U.S. EPA has developed an Air Quality Index to help explain air pollution levels to the general public. 8 hour average ozone concentrations of 85 to 104 ppbv are described as "Unhealthy for Sensitive Groups", 105 ppbv to 124 ppbv as "unhealthy" and 125 ppb to 404 ppb as "very unhealthy". The "very unhealthy" range for some other pollutants are: 355 µg m^{-3} - 424 µg m^{-3} for PM10; 15.5 ppm - 30.4ppm for CO and 0.65 ppm - 1.24 ppm for NO_2.

Premature Deaths Due to Cancer and Respiratory Disease

The Ontario Medical Association announced that smog is responsible for an estimated 9,500 premature deaths in the province each year.

A 20-year American Cancer Society study found that cumulative exposure also increases the likelihood of premature death from a respiratory disease, implying the 8-hour standard may be insufficient.

Smog and the Risk of Certain Birth Defects

A study examining 806 women who had babies with birth defects between 1997 and 2006, and 849 women who had healthy babies, found that smog in the San Joaquin Valley area of California was linked to two types of neural tube defects: spina bifida (a condition involving, among other manifestations, certain malformations of the spinal column), and anencephaly (the underdevelopment or absence of part or all of the brain, which if not fatal usually results in profound impairment).

Smog and Low Birth Weight

According to a study published in The Lancet, even a very small (5 µg) change in PM2.5 exposure was associated with an increase (18%) in risk of a low birth weight at delivery, and this relationship held even below the current accepted safe levels.

Areas Affected

Beijing air on a day after rain (left) and a smoggy day (right)

Smog can form in almost any climate where industries or cities release large amounts of air pollution, such as smoke or gases. However, it is worse during periods of warmer, sunnier weather when the upper air is warm enough to inhibit vertical circulation. It is especially prevalent in geologic basins encircled by hills or mountains. It often stays for an extended period of time over densely populated cities or urban areas, and can build up to dangerous levels.

Delhi, India

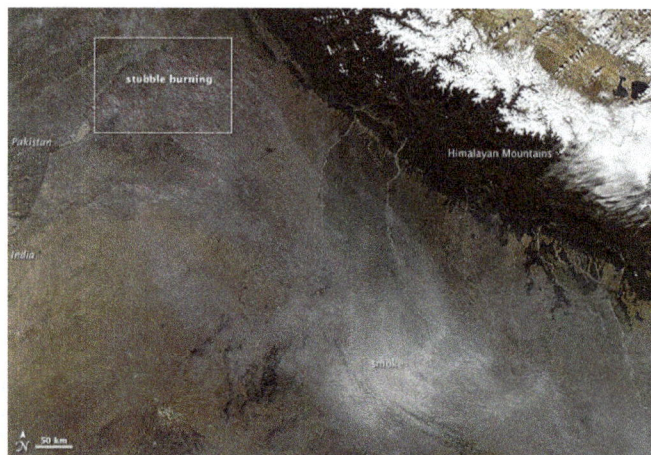

During the autumn and winter months, some 500 million tons of crop residue are burnt, and winds blow from India's north and northwest towards east. This aerial view shows India's annual crop burning, resulting in smoke and air pollution over Delhi and adjoining areas.

Delhi is the most polluted city in the world and according to one estimate, air pollution causes the death of about 10,500 people in Delhi every year. During 2013-14, peak levels of fine particulate matter (PM) in Delhi increased by about 44%, primarily due to high vehicular and industrial emissions, construction work and crop burning in adjoining states. Delhi has the highest level of the airborne particulate matter, PM2.5 considered most harmful to health, with 153 micrograms. Rising air pollution level has significantly increased lung-related ailments (especially asthma and lung cancer) among Delhi's children and women. The dense smog in Delhi during winter season results in major air and rail traffic disruptions every year. According to Indian meteorologists, the average maximum temperature in Delhi during winters has declined notably since 1998 due to rising air pollution.

Dense smog blankets Connaught Place, New Delhi.

Environmentalists have criticised the Delhi government for not doing enough to curb air pollution and to inform people about air quality issues. Most of Delhi's residents are unaware of alarming levels of air pollution in the city and the health risks associated with it. Since the mid-1990s, Delhi has undertaken some measures to curb air pollution – Delhi has the third highest quantity of trees among Indian cities and the Delhi Transport Corporation operates the world's largest fleet of environmentally friendly compressed natural gas (CNG) buses. In 1996, the Centre for Science and Environment (CSE) started a public interest litigation in the Supreme Court of India that ordered the conversion of Delhi's fleet of buses and taxis to run on CNG and banned the use of leaded petrol in 1998. In 2003, Delhi won the United States Department of Energy's first 'Clean Cities International Partner of the Year' award for its "bold efforts to curb air pollution and support alternative fuel initiatives". The Delhi Metro has also been credited for significantly reducing air pollutants in the city.

However, according to several authors, most of these gains have been lost, especially due to stubble burning, rise in market share of diesel cars and a considerable decline in bus ridership. According to CSE and System of Air Quality Weather Forecasting and Research (SAFAR), burning of agricultural waste in nearby Punjab, Haryana and Uttar Pradesh regions results in severe intensification of smog over Delhi. The state government of adjoining Uttar Pradesh is considering imposing a ban on crop burning to reduce pollution in Delhi NCR and an environmental panel has appealed to India's Supreme Court to impose a 30% cess on diesel cars.

United Kingdom

London

Victorian London was notorious for its thick smogs, or "pea-soupers", a fact that is often recreated (as here) to add an air of mystery to a period costume drama

In 1306, concerns over air pollution were sufficient for Edward I to (briefly) ban coal fires in London. In 1661, John Evelyn's *Fumifugium* suggested burning fragrant wood instead of mineral coal, which he believed would reduce coughing. The "Ballad of Gresham College" the same year describes how the smoke "does our lungs and spirits choke, Our hanging spoil, and rust our iron."

Severe episodes of smog continued in the 19th and 20th centuries, mainly in the winter, and were nicknamed "pea-soupers," from the phrase "as thick as pea soup." The Great Smog of 1952 darkened the streets of London and killed approximately 4,000 people in the short time of 4 days (a further 8,000 died from its effects in the following weeks and months). Initially a flu epidemic was blamed for the loss of life.

In 1956 the Clean Air Act started legally enforcing smokeless zones in the capital. There were areas where no soft coal was allowed to be burned in homes or in businesses, only coke, which produces no smoke. Because of the smokeless zones, reduced levels of sooty particulates eliminated the intense and persistent London smog.

It was after this that the great clean-up of London began. One by one, historical buildings which, during the previous two centuries had gradually completely blackened externally, had their stone facades cleaned and restored to their original appearance. Victorian buildings whose appearance changed dramatically after cleaning included the British Museum of Natural History. A more recent example was the Palace of Westminster, which was cleaned in the 1980s. A notable exception to the restoration trend was 10 Downing Street, whose bricks upon cleaning in the late 1950s proved to be naturally *yellow*; the smog-derived black colour of the façade was considered so iconic that the bricks were painted black to preserve the image. Smog caused by traffic pollution, however, does still occur in modern London.

Other Areas

Other areas of the United Kingdom were affected by smog, especially heavily industrialised areas.

The cities of Glasgow and Edinburgh, in Scotland, suffered smoke-laden fogs in 1909. Des Voeux, commonly credited with creating the "smog" moniker, presented a paper in 1911 to the Manchester Conference of the Smoke Abatement League of Great Britain about the fogs and resulting deaths.

One Birmingham resident described near black-out conditions in the 1900s before the Clean Air Act, with visibility so poor that cyclists had to dismount and walk in order to stay on the road.

Mexico City, Mexico

Situated in a valley, and relying heavily on automobiles, Mexico City often suffers from poor air quality.

Due to its location in a highland "bowl", cold air sinks down onto the urban area of Mexico City, trapping industrial and vehicle pollution underneath, and turning it into the most infamously smog-plagued city of Latin America. Within one generation, the city has changed from being known for some of the cleanest air of the world into one with some of the worst pollution, with pollutants like nitrogen dioxide being double or even triple international standards.

Photochemical smog over Mexico City. December 2010.

Santiago, Chile

Similar to Mexico City, the air pollution of Santiago valley, located between the Andes and the Chilean Coast Range, turn it into the most infamously smog-plagued city of South America. Other aggravates of the situation reside in its high latitude (31 degrees South) and dry weather during most of the year.

Tehran, Iran

In December 2005, schools and public offices had to close in Tehran, Iran and 1600 people were taken to hospital, in a severe smog blamed largely on unfiltered car exhaust.

United States

A NASA astronaut photograph of a smog layer over central New York.

View of smog south from Los Angeles City Hall, September 2011.

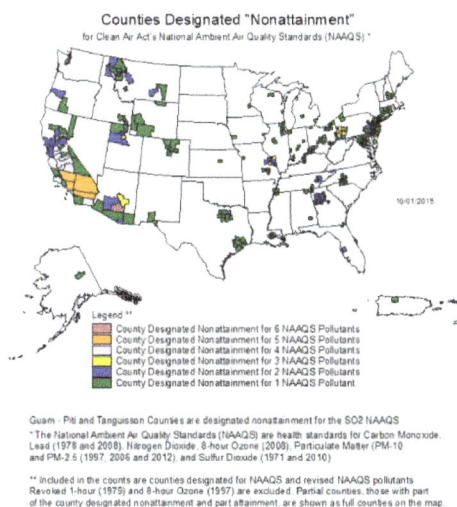

Counties in the United States where one or more National Ambient Air Quality Standards are not met, as of October 2015.

Smog was brought to the attention of the general US public in 1933 with the publication of the book "Stop That Smoke", by Henry Obermeyer, a New York public utility official, in which he pointed out the effect on human life and even the destruction of 3,000 acres (12 km²) of a farmer's spinach crop. Since then, the United States Environmental Protection Agency has designated over 300 U.S. counties to be non-attainment areas for one or more pollutants tracked as part of the National Ambient Air Quality Standards. These areas are largely clustered around large metropolitan areas, with the largest contiguous non-attainment zones in California and the Northeast. Various U.S. and Canadian government agencies collaborate to produce real-time air quality maps and forecasts.

Los Angeles and the San Joaquin Valley

Because of their locations in low basins surrounded by mountains, Los Angeles and the San Joaquin Valley are notorious for their smog. The millions of vehicles in these regions combined with the additional effects of the San Francisco Bay and Los Angeles/Long Beach port complexes frequently contribute to further air pollution.

Los Angeles in particular is strongly predisposed to accumulation of smog, because of peculiarities of its geography and weather patterns. Los Angeles is situated in a flat basin with ocean on one side and mountain ranges on three sides. A nearby cold ocean current depresses surface air temperatures in the area, resulting in an inversion layer: a phenomenon where air temperature increases, instead of decreasing, with altitude, suppressing thermals and restricting vertical convection. All taken together, this results in a relatively thin, enclosed layer of air above the city that cannot easily escape out of the basin and tends to accumulate pollution.

Though Los Angeles was one of the best known cities suffering from transportation smog for much of the 20th century, so much so that it was sometimes said that *Los Angeles* was a synonym for *smog*, strict regulations by government agencies overseeing this problem, including tight restrictions on allowed emissions levels for all new cars sold in California and mandatory regular emission tests of older vehicles, resulted in significant improvements in air quality. For example, air concentrations of volatile organic compounds declined by a factor of 50 between 1962 and 2012. Concentrations of air pollutants such as nitrous oxides and ozone declined by 70% to 80% over the same period of time.

Major incidents in the US

- 1943, July 26, Los Angeles, California: A smog so sudden and severe that "Los Angeles residents believe the Japanese are attacking them with chemical warfare."

- 1948, October 30–31, Donora, Pennsylvania: 20 died, 600 hospitalized, thousands more stricken. Lawsuits were not settled until 1951.

- 1966, November 24, New York City, New York: Smog kills at least 169 people.

Ulaanbaatar, Mongolia

In the late 1990s, massive immigration to Ulaanbaatar from the countryside began. An estimated 150,000 households, mainly living in traditional Mongolian gers on the outskirts of Ulaanbaatar, burn wood and coal (some poor families burn even car tires and trash) to heat themselves during the harsh winter, which lasts from October to April, since these outskirts are not connected to the city's central heating system. A temporary solution to decrease smog was proposed in the form of stoves with improved efficiency, although with no visible results. Coal-fired ger stoves release high levels of ash and other particulate matter (PM). When inhaled, these particles can settle in the lungs and respiratory tract and cause health problems. At two to 10 times above Mongolian and international air quality standards, Ulaanbaatar's PM rates are among the worst in the world, according to a December 2009 World Bank report. The Asian Development Bank (ADB) estimates that health costs related to this air pollution account for as much as 4 percent of Mongolia's GDP.

Southeast Asia

Singapore's Downtown Core on 7 October 2006, when it was affected by forest fires in Sumatra, Indonesia

Smog is a regular problem in Southeast Asia caused by land and forest fires in Indonesia, especially Sumatra and Kalimantan, although the term haze is preferred in describing the problem. Farmers and plantation owners are usually responsible for the fires, which they use to clear tracts of land for further plantings. Those fires mainly affect Brunei, Indonesia, Philippines, Malaysia, Singapore and Thailand, and occasionally Guam and Saipan. The economic losses of the fires in 1997 have been estimated at more than US$9 billion. This includes damages in agriculture production, destruction of forest lands, health, transportation, tourism, and other economic endeavours. Not included are social, environmental, and psychological problems and long-term health effects. The second-latest bout of haze to occur in Malaysia, Singapore and the Malacca Straits is in October 2006, and was caused by smoke from fires in Indonesia being blown across the Straits of Malacca by south-westerly winds. A similar haze has occurred in June 2013, with the PSI setting a new record in Singapore on June 21 at 12pm with a reading of 401, which is in the "Hazardous" range.

The Association of Southeast Asian Nations (ASEAN) reacted. In 2002, the Agreement on Transboundary Haze Pollution was signed between all ASEAN nations. ASEAN formed a Regional Haze Action Plan (RHAP) and established a co-ordination and support unit (CSU). RHAP, with the help of Canada, established a monitoring and warning system for forest/vegetation fires and implemented a Fire Danger Rating System (FDRS). The Malaysian Meteorological Department (MMD) has issued a daily rating of fire danger since September 2003. Indonesia has been ineffective at enforcing legal policies on errant farmers.

Pollution Index

Smog in São Paulo, Brazil

The severity of smog is often measured using automated optical instruments such as Nephelometers, as haze is associated with visibility and traffic control in ports. Haze however can also be an indication of poor air quality though this is often better reflected using accurate purpose built air indexes such as the American Air Quality Index, the Malaysian API (Air Pollution Index) and the Singaporean Pollutant Standards Index.

In hazy conditions, it is likely that the index will report the suspended particulate level. The disclosure of the responsible pollutant is mandated in some jurisdictions.

The Malaysian API does not have a capped value; hence its most hazardous readings can go above 500. Above 500, a state of emergency is declared in the affected area. Usually, this means that non-essential government services are suspended, and all ports in the affected area are closed. There may also be prohibitions on private sector commercial and industrial activities in the affected area excluding the food sector. So far, state of emergency rulings due to hazardous API levels were applied to the Malaysian towns of Port Klang, Kuala Selangor and the state of Sarawak during the 2005 Malaysian haze and the 1997 Southeast Asian haze.

Cultural References

Claude Monet made several trips to London between 1899 and 1901, during which he painted views of the Thames and Houses of Parliament which show the sun struggling to shine through London's smog-laden atmosphere.

- The London "pea-soupers" earned the capital the nickname of "The Smoke". Similarly, Edinburgh was known as "Auld Reekie". The smogs feature in many London novels as a motif indicating hidden danger or a mystery, perhaps most overtly in Margery Allingham's *The Tiger in the Smoke* (1952), but also in Dickens's *Bleak House* (1852) and T.S. Eliot's "The Love Song of J. Alfred Prufrock."

- The 1970 made-for-TV movie *A Clear and Present Danger* was one of the first American television network entertainment programs to warn about the problem of smog and air pollution, as it dramatized a man's efforts toward clean air after emphysema killed his friend.

- The history of smog in LA is detailed in *Smogtown* by Chip Jacobs and William J. Kelly.

Thermal Pollution

The Brayton Point Power Station in Massachusetts discharges heated water to Mount Hope Bay.

Thermal pollution is the degradation of water quality by any process that changes ambient water temperature. A common cause of thermal pollution is the use of water as a coolant by power plants and industrial manufacturers. When water used as a coolant is returned to the natural environment at a higher temperature, the change in temperature decreases oxygen supply and affects ecosystem composition. Fish and other organisms adapted to particular temperature range can be killed by an abrupt change in water temperature (either a rapid increase or decrease) known as "thermal shock."

Urban runoff—stormwater discharged to surface waters from roads and parking lots—can also be a source of elevated water temperatures.

Ecological Effects

Potrero Generating Station discharged heated water into San Francisco Bay. The plant was closed in 2011.

Warm Water

Elevated temperature typically decreases the level of dissolved oxygen of water. This can harm aquatic animals such as fish, amphibians and other aquatic organisms. Thermal pollution may also increase the metabolic rate of aquatic animals, as enzyme activity, resulting in these organisms consuming more food in a shorter time than if their environment were not changed. An increased metabolic rate may result in fewer resources; the more adapted organisms moving in may have an advantage over organisms that are not used to the warmer temperature. As a result, food chains of the old and new environments may be compromised. Some fish species will avoid stream segments or coastal areas adjacent to a thermal discharge. Biodiversity can be decreased as a result.

High temperature limits oxygen dispersion into deeper waters, contributing to anaerobic conditions. This can lead to increased bacteria levels when there is ample food supply. Many aquatic species will fail to reproduce at elevated temperatures.

Primary producers (e.g. plants, cyanobacteria) are affected by warm water because higher water temperature increases plant growth rates, resulting in a shorter lifespan and species overpopulation. This can cause an algae bloom which reduces oxygen levels.

Temperature changes of even one to two degrees Celsius can cause significant changes in organism metabolism and other adverse cellular biology effects. Principal adverse changes can include rendering cell walls less permeable to necessary osmosis, coagulation of cell proteins, and alteration of enzyme metabolism. These cellular level effects can adversely affect mortality and reproduction.

A large increase in temperature can lead to the denaturing of life-supporting enzymes by breaking down hydrogen- and disulphide bonds within the quaternary structure of the enzymes. Decreased enzyme activity in aquatic organisms can cause problems such as the inability to break down lipids, which leads to malnutrition.

In limited cases, warm water has little deleterious effect and may even lead to improved function of the receiving aquatic ecosystem. This phenomenon is seen especially in seasonal waters and is known as thermal enrichment. An extreme case is derived from the aggregational habits of the manatee, which often uses power plant discharge sites during winter. Projections suggest that manatee populations would decline upon the removal of these discharges.

Cold Water

Releases of unnaturally cold water from reservoirs can dramatically change the fish and macroinvertebrate fauna of rivers, and reduce river productivity. In Australia, where many rivers have warmer temperature regimes, native fish species have been eliminated, and macroinvertebrate fauna have been drastically altered. This may be mitigated by designing the dam to release warmer surface waters instead of the colder water at the bottom of the reservoir.

Thermal Shock

When a power plant first opens or shuts down for repair or other causes, fish and other organisms adapted to particular temperature range can be killed by the abrupt change in water temperature, either an increase or decrease, known as "thermal shock."

Sources and Control of Thermal Pollution

Cooling tower at Gustav Knepper Power Station, Dortmund, Germany

Industrial Wastewater

In the United States, about 75 to 82 percent of thermal pollution is generated by power plants. The remainder is from industrial sources such as petroleum refineries, pulp and paper mills, chemical plants, steel mills and smelters. Heated water from these sources may be controlled with:

cooling ponds, man-made bodies of water designed for cooling by evaporation, convection, and radiation

cooling towers, which transfer waste heat to the atmosphere through evaporation and/or heat transfer

cogeneration, a process where waste heat is recycled for domestic and/or industrial heating purposes.

Some facilities use once-through cooling (OTC) systems which do not reduce temperature as effectively as the above systems. For example, the Potrero Generating Station in San Francisco (closed in 2011), used OTC and discharged water to San Francisco Bay approximately 10 °C (20 °F) above the ambient bay temperature.

A bioretention cell for treating urban runoff in California

Urban Runoff

During warm weather, urban runoff can have significant thermal impacts on small streams, as stormwater passes over hot parking lots, roads and sidewalks. Stormwater management facilities that absorb runoff or direct it into groundwater, such as bioretention systems and infiltration basins, can reduce these thermal effects. These and related systems for managing runoff are components of an expanding urban design approach commonly called green infrastructure.

Retention basins (stormwater ponds) tend to be less effective at reducing runoff temperature, as the water may be heated by the sun before being discharged to a receiving stream.

References

- UK Royal Society (September 2009). "Geoengineering the climate: science, governance and uncertainty" (PDF). London: UK Royal Society. ISBN 978-0-85403-773-5 , RS Policy document 10/09. Report website.

- WBGU (2006). Special Report: The Future Oceans – Warming Up, Rising High, Turning Sour (PDF). Berlin, Germany: WBGU. ISBN 3-936191-14-X . Report website.

- Seinfeld, John H.; Pandis, Spyros N (1998). Atmospheric Chemistry and Physics — From Air Pollution to Climate Change. John Wiley and Sons, Inc. ISBN 978-0-471-17816-3

- Berresheim, H.; Wine, P.H. and Davies D.D. (1995). "Sulfur in the Atmosphere". In Composition, Chemistry and Climate of the Atmosphere, ed. H.B. Singh. Van Nostrand Rheingold ISBN 0-442-01264-0

- Miller, Jr., George Tyler (2002). Living in the Environment: Principles, Connections, and Solutions (12th Edition). Belmont: The Thomson Corporation. p. 423. ISBN 0-534-37697-5.

- Buntin, John (2009). L.A. Noir: The Struggle for the Soul of America's Most Seductive City. New York: Harmony Books. p. 108. ISBN 9780307352071. OCLC 431334523. Retrieved 12 October 2014.

- Jacobs, Chip; Kelly, William J. (4 October 2009). Smogtown, The Lung-Burning History of Pollution in Los Angeles. Overlook Press. ISBN 978-1-58567-860-0.

- Kennish, Michael J. (1992). Ecology of Estuaries: Anthropogenic Effects. Marine Science Series. Boca Raton, FL: CRC Press. ISBN 978-0-8493-8041-9.

- IPCC TAR WG3 (2001). Metz, B.; Davidson, O.; Swart, R.; Pan, J., eds. Climate Change 2001: Mitigation. Contribution of Working Group III to the Third Assessment Report of the Intergovernmental Panel on Climate Change. Cambridge University Press. ISBN 0-521-80769-7.

- National Research Council (2010). America's Climate Choices: Panel on Advancing the Science of Climate Change;. Washington, D.C.: The National Academies Press. ISBN 0-309-14588-0.

- Roan, Sharon (1989) Ozone crisis: The 15-year evolution of a sudden global emergency, New York: Wiley, p. 56 ISBN 0-471-52823-4

- Causes and Effects of Stratospheric Ozone Reduction: An Update. National Research Council. 1982. ISBN 0-309-03248-2.

- Kump, Lee R.; Kasting, James F.; Crane, Robert G. (2003). The Earth System (2nd ed.). Upper Saddle River: Prentice Hall. pp. 162–164. ISBN 0-613-91814-2.

- Gardner, Sarah (14 July 2014). "LA Smog: the battle against air pollution". Marketplace.org. American Public Media. Retrieved 6 November 2015.

- Marziali, Carl (4 March 2015). "L.A.'s Environmental Success Story: Cleaner Air, Healthier Kids". USC News. Retrieved 16 March 2015.

- Tracton, Steve (December 20, 2012). "The Killer London Smog Event of December 1952: A Reminder of Deadly Smog Events in the US". The Washington Post. Retrieved February 25, 2015.

- "Brayton Point Station: Final NPDES Permit". NPDES Permits in New England. U.S. Environmental Protection Agency (EPA), Boston, MA. 2014. Retrieved 2015-04-13.

- "Twenty Questions and Answers About the Ozone Layer". Scientific Assessment of Ozone Depletion: 2010 (PDF). World Meteorological Organization. 2011. Retrieved March 13, 2015.

- "Lanzan la primera de las "Propuestas Greenpeace": la heladera "Greenfreeze" | Greenpeace Argentina". Greenpeace.org. Retrieved September 27, 2015.

Pollution: Challenges and Concerns

As time passes and pollution levels remain unchecked, vulnerable areas of the Earth suffer from such accumulation of waste and the disruption of the natural cycles of the ecosystem. The ozone layer, the Arctic and increasing intensity of natural calamities are some of the aspects of rampant pollution. This chapter discusses the challenges of pollution in a critical manner providing key analysis to the subject matter.

Arctic Sea Ice Decline

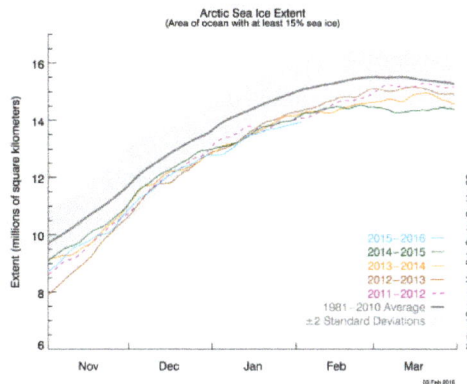

Arctic sea ice extent as of February 3, 2016. January Arctic sea ice extent was the lowest in the satellite record. credit: NSIDC

Arctic sea ice decline is the sea ice loss observed in recent decades in the Arctic Ocean. The Intergovernmental Panel on Climate Change (IPCC) Fourth Assessment Report states that greenhouse gas forcing is largely, but not wholly, responsible for the decline in Arctic sea ice extent. A study from 2011 found the decline to be "faster than forecasted" by model simulations. The IPCC Fifth Assessment Report concluded with high confidence that sea ice continues to decrease in extent, and that there is robust evidence for the downward trend in Arctic summer sea ice extent since 1979. It has been established that the region is at its warmest for at least 40,000 years and the Arctic-wide melt season has lengthened at a rate of 5 days per decade (from 1979 to 2013), dominated by a later autumn freezeup. Sea ice changes have been identified as a mechanism for polar amplification.

Definitions

The Arctic sea ice minimum is the day in a given year when Arctic sea ice reaches its smallest extent. It occurs at the end of the summer melting season, normally during September. Sea ice maximum is the day of a year when Arctic sea ice reaches its largest extent near the end of the Arctic cold season, normally during March. Typical data visualizations for Arctic sea ice include average monthly measurements or graphs for the annual minimum or maximum extent, as shown in the images to the right.

Observation

Observation with satellites show that Arctic sea ice area, extent, and volume have been in decline for a few decades. Sometime during the 21st century, sea ice may effectively cease to exist during the summer. Sea ice extent is defined as the area with at least 15% ice cover. The amount of multi-year sea ice in the Arctic has declined considerably in recent decades. In 1988, ice that was at least 4 years old accounted for 26% of the Arctic's sea ice. By 2013, ice that age was only 7% of all Arctic sea ice.

Scientists recently measured sixteen-foot (five-meter) wave heights during a storm in the Beaufort Sea in mid-August until late October 2012. This is a new phenomenon for the region, since a permanent sea ice cover normally prevents wave formation. Wave action breaks up sea ice, and thus could become a feedback mechanism, driving sea ice decline.

For January 2016, the satellite based data showed the lowest overall Arctic sea ice extent of any January since records begun in 1979. Bob Henson from Wunderground noted:

Hand in hand with the skimpy ice cover, temperatures across the Arctic have been extraordinarily warm for midwinter. Just before New Year's, a slug of mild air pushed temperatures above freezing to within 200 miles of the North Pole. That warm pulse quickly dissipated, but it was followed by a series of intense North Atlantic cyclones that sent very mild air poleward, in tandem with a strongly negative Arctic Oscillation during the first three weeks of the month.

Ice-free Summer

As ice melts, the liquid water collects in depressions on the surface and deepens them, forming these melt ponds in the Arctic. These fresh water ponds are separated from the salty sea below and around it, until breaks in the ice merge the two.

An "ice-free" Arctic Ocean is often defined as "having less than 1 million square kilometers of sea ice", because it is very difficult to melt the thick ice around the Canadian Arctic Archipelago. The IPCC AR5 defines "nearly ice-free conditions" as sea ice extent less than 10^6 km^2 for at least five consecutive years.

Many scientists have attempted to estimate when the Arctic will be "ice-free". They have noted that climate model predictions have tended to be overly conservative regarding sea ice decline. A 2013 paper suggested that models commonly underestimate the solar radiation absorption characteristics of wildfire soot. A 2006 paper predicted "near ice-free September conditions by 2040". Overland & Wang (2009) predicted that there would be an ice-free Arctic in the summer by 2037. The same year Boé et al. found that the Arctic will probably be ice-free in September before the end of the 21st century. A follow-up study concluded with the possibility of major sea ice loss within a decade or two. The IPCC AR5 (for at least one scenario) estimates an ice-free summer might occur around 2050. The Third U.S. National Climate Assessment (NCA), released May 6, 2014, reports that the Arctic Ocean is expected to be ice free in summer before mid-century. Simulations by global climate models generally map well to this seasonal pattern of observed Arctic sea ice loss. Models that best match historical trends project a nearly ice-free Arctic in the summer by the 2030s. However, these models do tend to underestimate the rate of sea ice loss since 2007. A 2010 paper suggests that the Arctic Ocean will be ice-free sooner than global climate models predict. They chart the summer of 2016 as ice-free, but show a possible date range out to 2020. This assessment was reported in the press as "US Navy predicts summer ice free Arctic by 2016"

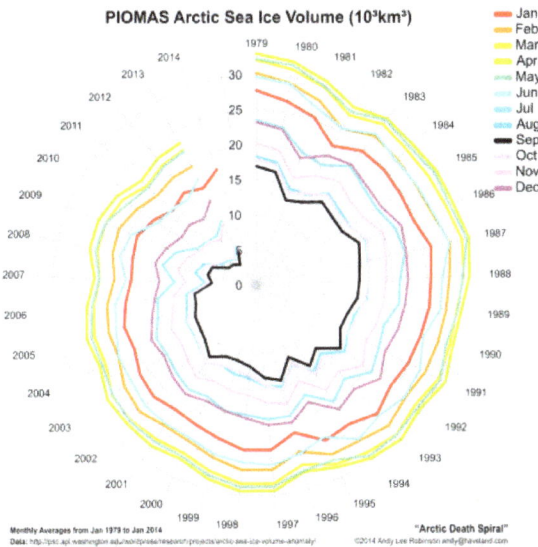

Monthly averages from January 1979 - January 2014. Data source via the Polar Science Center (University of Washington). Data visualisation by Andy Lee Robinson.

Tipping Point

There is an ongoing debate if the Arctic Ocean has already passed a "tipping point", defined as a threshold for abrupt and irreversible change. A 2013 study identified an abrupt transition to increased seasonal ice cover variability in 2007, which has persisted in following years. The researchers made a distinction between a bifurcation and a non-bifurcation `tipping point'. The *IPCC AR5* report stated with medium confidence that precise levels of climate change sufficient to trigger a tipping point remain uncertain, and that the risk associated with crossing multiple tipping points increases with rising temperature.

Implications

Implications which arise from lesser ocean surface covered with sea-ice include the ice-albedo feedback or warmer sea surface temperatures which increase ocean heat content, which in turn changes evaporation patterns and the polar vortex.

Atmospheric Chemistry

Melting of sea ice releases molecular chlorine, which reacts with sunlight to produce chlorine atoms. Because chlorine atoms are highly reactive, they can expedite the degradation of methane and tropospheric ozone and the oxidation of mercury to more toxic forms. Cracks in sea ice are causing ozone and mercury uptake in the surrounding environment.

A 2015 study concluded that Arctic sea ice decline accelerates methane emissions from the Arctic tundra. One of the study researchers noted, "The expectation is that with further sea ice decline, temperatures in the Arctic will continue to rise, and so will methane emissions from northern wetlands."

Atmospheric Regime

A link has been proposed between reduced Barents-Kara sea ice and cold winter extremes over northern continents. Model simulation suggest diminished Arctic sea ice may have been a contributing driver of recent wet summers over northern Europe, because of a weakened jet stream, which dives further south. Extreme summer weather in northern mid-latitudes has been linked to a vanishing cryosphere. Evidence suggest that the continued loss of Arctic sea-ice and snow cover may influence weather at lower latitudes. Correlations have been identified between high-latitude cryosphere changes, hemispheric wind patterns and mid-latitude extreme weather events for the Northern Hemisphere. A study from 2004, connected the disappearing sea ice with a reduction of available water in the American west.

Based on effects of Arctic amplification (warming) and ice loss, a study in 2015 concluded that highly amplified jet-stream patterns are occurring more frequently in the past two decades, and that such patterns can not be tied to certain seasons. Additionally it was found that these jet-stream patterns often lead to persistent weather patterns that result in extreme weather events. Hence, continued heat trapping emissions favour increased formation of extreme events caused by prolonged weather conditions.

Plant and Animal Life

Sea ice decline has been linked to boreal forest decline in North America and is assumed to culminate with an intensifying wildfire regime in this region. The annual net primary production of the Eastern Bering Sea was enhanced by 40–50% through phytoplankton blooms during warm years of early sea ice retreat.

Polar bears are turning to alternate food sources because Arctic sea ice melts earlier and freezes later each year. As a result, they have less time to hunt their historically preferred prey of seal pups, and must spend more time on land and hunt other animals. As a result, the diet is less nutritional, which leads to reduced body size and reproduction, thus indicating population decline in polar bears.

Emission Intensity

An emission intensity is the average emission rate of a given pollutant from a given source relative to the intensity of a specific activity; for example grams of carbon dioxide released per megajoule of energy produced, or the ratio of greenhouse gas emissions produced to gross domestic product (GDP). Emission intensities are used to derive estimates of air pollutant or greenhouse gas emissions based on the amount of fuel combusted, the number of animals in animal husbandry, on industrial production levels, distances traveled or similar activity data. Emission intensities may also be used to compare the environmental impact of different fuels or activities. The related terms emission factor and carbon intensity are often used interchangeably, but "factors" exclude aggregate activities such as GDP, and "carbon" excludes other pollutants. One commonly used figure is carbon intensity per kilowatt, or CIPK, which is used to compare different sources of electrical power.

An air pollution emission source

Estimating Emissions

Emission factors assume a linear relation between the intensity of the activity and the emission resulting from this activity:

$Emission_{pollutant} = Activity * Emission\ Factor_{pollutant}$

Intensities are also used in projecting possible future scenarios such as those used in the IPCC assessments, along with projected future changes in population, economic activity and energy technologies. The interrelations of these variables is treated under the so-called Kaya identity.

The level of uncertainty of the resulting estimates depends significantly on the source category and the pollutant. Some examples:

- Carbon dioxide (CO_2) emissions from the combustion of fuel can be estimated with a high degree of certainty regardless of how the fuel is used as these emissions depend almost exclusively on the carbon content of the fuel, which is generally known with a high degree of precision. The same is true for sulphur dioxide (SO_2), since sulphur contents of fuels are also generally well known. Both carbon and sulphur are almost completey oxidized during combustion and all carbon and sulphur atoms in the fuel will be present in the flue gases as CO_2 and SO_2 respectively.

- In contrast, the levels of other air pollutants and non-CO_2 greenhouse gas emissions from combustion depend on the precise technology applied when fuel is combusted. These emissions are basically caused by either incomplete combustion of a small fraction of the fuel (carbon monoxide, methane, non-methane volatile organic compounds) or by complicated chemical and physical processes during the combustion and in the smoke stack or tailpipe. Examples of these are particulates, NO_x, a mixture of nitric oxide, NO, and nitrogen dioxide, NO_2).

- Nitrous oxide (N_2O) emissions from agricultural soils are highly uncertain because they depend very much on both the exact conditions of the soil, the application of fertilizers and meteorological conditions.

Published Data on Energy Sources Emission Intensity Per Unit of Energy Generated

A literature review of numerous total life cycle energy sources CO_2 emissions per unit of electricity generated, conducted by the Intergovernmental Panel on Climate Change in 2011, found that the CO_2 emission value, that fell within the 50th percentile of all total life cycle emissions studies were as follows.

Lifecycle greenhouse gas emissions by electricity source.		
Technology	**Description**	**50th percentile (g CO_2-eq/kWh$_e$)**
Hydroelectric	reservoir	4
Wind	onshore	12
Nuclear	various generation II reactor types	16
Biomass	Various	18
Solar thermal	parabolic trough	22
Geothermal	hot dry rock	45
Solar PV	Polycrystalline silicon	46
Natural gas	various combined cycle turbines without scrubbing	469
Coal	various generator types without scrubbing	1001

Emission factors of common fuels			
Fuel/Resource	Thermal g(CO_2-eq)/MJ_{th}	Energy Intensity (min & max estimate) $W \cdot h_{th}$/$W \cdot h_e$	Electric (min & max estimate) g(CO_2-eq)/$kW \cdot h_e$
wood	115		
Peat	106 110		
Coal	B:91.50–91.72 Br:94.33 88	B:2.62–2.85 Br:3.46 3.01	B:863–941 Br:1,175 955
Oil	73	3.40	893
Natural gas	cc:68.20 oc:68.40 51	cc:2.35 (2.20 – 2.57) oc:3.05 (2.81 – 3.46)	cc:577 (491–655) oc:751 (627–891) 599
Geothermal Power	3~		T_L0–1 T_H91–122
Uranium Nuclear power		W_L0.18 (0.16~0.40) W_H0.20 (0.18~0.35)	W_L60 (10~130) W_H65 (10~120)
Hydroelectricity		0.046 (0.020 – 0.137)	15 (6.5 – 44)
Conc. Solar Pwr			40±15#
Photovoltaics		0.33 (0.16 – 0.67)	106 (53–217)
Wind power		0.066 (0.041 – 0.12)	21 (13–40)

Note: 3.6 MJ = megajoule(s) == 1 kW·h = kilowatt-hour(s), thus 1 g/MJ = 3.6 g/kW·h. Legend: B = Black coal (supercritical)–(new subcritical), Br = Brown coal (new subcritical), cc = combined cycle, oc = open cycle, T_L = low-temperature/closed-circuit (geothermal doublet), T_H = high-temperature/open-circuit, W_L = Light Water Reactors, W_H = Heavy Water Reactors, #Educated estimate.

Carbon Intensity of Regions

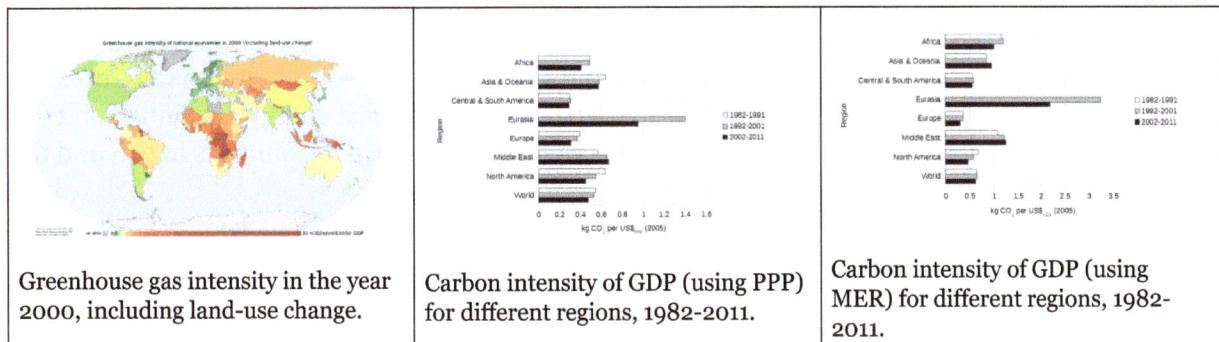

Greenhouse gas intensity in the year 2000, including land-use change.	Carbon intensity of GDP (using PPP) for different regions, 1982-2011.	Carbon intensity of GDP (using MER) for different regions, 1982-2011.

The following tables show carbon intensity of GDP in market exchange rates (MER) and purchasing power parities (PPP). Units are metric tons of carbon dioxide per thousand year 2005 US dollars. Data are taken from the US Energy Information Administration. Annual data between 1980 and 2009 are averaged over three decades: 1980-89, 1990–99, and 2000–09.

Carbon intensity of GDP, measured in MER			
	1980-89	**1990-99**	**2000-09**
Africa	1.13149	1.20702	1.03995
Asia & Oceania	0.86256	0.83015	0.91721
Central & South America	0.55840	0.57278	0.56015
Eurasia	NA	3.31786	2.36849
Europe	0.36840	0.37245	0.30975
Middle East	0.98779	1.21475	1.22310
North America	0.69381	0.58681	0.48160
World	0.62170	0.66120	0.60725

Carbon intensity of GDP, measured in PPP			
	1980-89	**1990-99**	**2000-09**
Africa	0.48844	0.50215	0.43067
Asia & Oceania	0.66187	0.59249	0.57356
Central & South America	0.30095	0.30740	0.30185
Eurasia	NA	1.43161	1.02797
Europe	0.40413	0.38897	0.32077
Middle East	0.51641	0.65690	0.65723
North America	0.66743	0.56634	0.46509
World	0.54495	0.54868	0.48058

In 2009 CO_2 intensity of GDP in the OECD countries reduced by 2.9% and amounted to 0.33 $kCO_2/\$05p$ in the OECD countries. ("$05p" = 2005 US dollars, using purchasing power parities). The USA posted a higher ratio of 0.41 $kCO_2/\$05p$ while Europe showed the largest drop in CO_2 intensity compared to the previous year (−3.7%). CO_2 intensity continued to be roughly higher in non-OECD countries. Despite a slight improvement, China continued to post a high CO_2 intensity (0.81 $kCO_2/\$05p$). CO_2 intensity in Asia rose by 2% during 2009 since energy consumption continued to develop at a strong pace. Important ratios were also observed in countries in CIS and the Middle East.

Carbon Intensity in Europe

Total CO_2 emissions from energy use were 5% below their 1990 level in 2007. Over the period 1990–2007, CO_2 emissions from energy use have decreased on average by 0.3%/year although the economic activity (GDP) increased by 2.3%/year. After dropping until 1994 (−1.6%/year), the CO_2 emissions have increased steadily (0.4%/year on average) until 2003 and decreased slowly again since (on average by 0.6%/year). Total CO_2 emissions per capita decreased from 8.7 t in 1990 to 7.8 t in 2007, that is to say a decrease by 10%. Almost 40% of the reduction in CO_2 intensity is due to increased use of energy carriers with lower emission factors Total CO_2 emissions per unit of GDP, the "CO_2 intensity", decreased more rapidly than energy intensity: by 2.3%/year and 1.4%/year, respectively, on average between 1990 and 2007.

Carbon Intensity in 2012

Commodity Exchange Bratislava (CEB) has calculated carbon intensity for Voluntary Emissions Reduction projects carbon intensity in 2012 to be 0,343 tn/MWh.

Emission Factors for Greenhouse Gas Inventory Reporting

One of the most important uses of emission factors is for the reporting of national greenhouse gas inventories under the United Nations Framework Convention on Climate Change (UNFCCC). The so-called Annex I Parties to the UNFCCC have to annually report their national total emissions of greenhouse gases in a formalized reporting format, defining the source categories and fuels that must be included.

The UNFCCC has accepted the Revised 1996 IPCC Guidelines for National Greenhouse Gas Inventories, developed and published by the Intergovernmental Panel on Climate Change (IPCC) as the emission estimation methods that must be used by the parties to the convention to ensure transparency, completeness, consistency, comparability and accuracy of the national greenhouse gas inventories. These IPCC Guidelines are the primary source for default emission factors. Recently IPCC has published the 2006 IPCC Guidelines for National Greenhouse Gas Inventories. These and many more greenhouse gas emission factors can be found on IPCC's Emission Factor Database. Commercially applicable organisational greenhouse gas emission factors can be found on the search engine, EmissionFactors.com.

Particularly for non-CO_2 emissions, there is often a high degree of uncertainty associated with these emission factors when applied to individual countries. In general, the use of country-specific emission factors would provide more accurate estimates of emissions than the use of the default emission factors. According to the IPCC, if an activity is a major source of emissions for a country ('key source'), it is 'good practice' to develop a country-specific emission factor for that activity.

Emission Factors for Air Pollutant Inventory Reporting

National Air Pollution Emission Inventories are required annually under the provisions of the UNECE Convention on Long-Range Transboundary Air Pollution (LRTAP). Emission estimation methods and the associated emission factors for air pollutants have been developed by the EMEP Task Force on Emission Inventories and Projections (TFEIP) and are published in the EMEP/CORINAIR Emission Inventory Guidebook.

Intensity Targets

The U.S. plans to cut carbon intensity per dollar of GDP by 18% by 2012. This has been criticised by the World Resources Institute as this approach does not ensure absolute reductions if GDP grows faster than intensity declines.

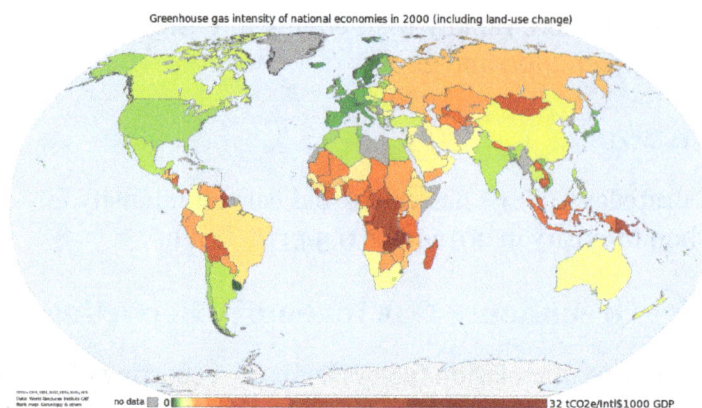

Greenhouse gas intensity in 2000 including land-use change

From 1990 to 2000, the carbon intensity of the U.S. economy declined by 17%, yet total emissions increased by 14%. In 2002, the U.S. National Environmental trust labelled carbon intensity, "a bookkeeping trick which allows the administration to do nothing about global warming while unsafe levels of emissions continue to rise."

Sources of Emission Factors

Greenhouse Gases

- 2006 IPCC Guidelines for National Greenhouse Gas Inventories
- Revised 1996 IPCC Guidelines for National Greenhouse Gas Inventories (reference manual).
- IPCC Emission Factor Database
- National Inventory Report: Greenhouse Gas Sources and Sinks in Canada.
- United Kingdom's emission factor database.

Air Pollutants

- AP 42, Compilation of Air Pollutant Emission Factors US Environmental Protection Agency
- EMEP/CORIMAIR 2007 Emission Inventory Guidebook.
- Fugitive emissions leaks from ethylene and other chemical plants.

Effects of Global Warming

The effects of global warming are the environmental and social changes caused (directly or indirectly) by human emissions of greenhouse gases. There is a scientific consensus that climate change is occurring, and that human activities are the primary driver. Many impacts of climate change have already been observed, including glacier retreat, changes in the timing of seasonal events (e.g., earlier flowering of plants), and changes in agricultural productivity.

Increase in global mean temperature, relative to pre-industrial levels

0   1   2   3   4   5   6   7   8   9 °F

Physical	Increase in risk associated with some extreme weather events: *Moderate Large*
	Other effects include global mean sea level rise and ocean acidification. Global warming could be irreversible for several millennia.
Ecological	Climate change already poses a significant risk to vulnerable systems, e.g., Arctic ecosystems and coral reefs. Risks to these systems are large even with small temperature increases.
	Risk of widespread extinctions: *Small Moderate Large*
Social	Sectors affected include food security, water resources and human health. Impacts will be uneven within and across different countries. Climate change increases the risk of many negative impacts, but there will be some positive effects.
	Africa: Risks associated with reduced crop productivity: *Low to moderate Moderate to high Very high*
	North America: Risks associated with urban flooding in riverine and coastal areas: *Low to moderate Moderate High*
Abrupt and large-scale changes	Climate change can lead to abrupt and large-scale changes in natural and human systems. The risk of these changes increases with temperature.
	Late-summer Arctic sea ice extent has already substantially declined, and is expected to decrease further at low temperatures.
	Sustained warming could lead to the near-complete loss of the Greenland ice sheet over a millennium or more, leading to global sea level rise of 7 m.

0   ▲ 1   2   3   4   5 °C

Recent temperatures
(2003 to 2012)

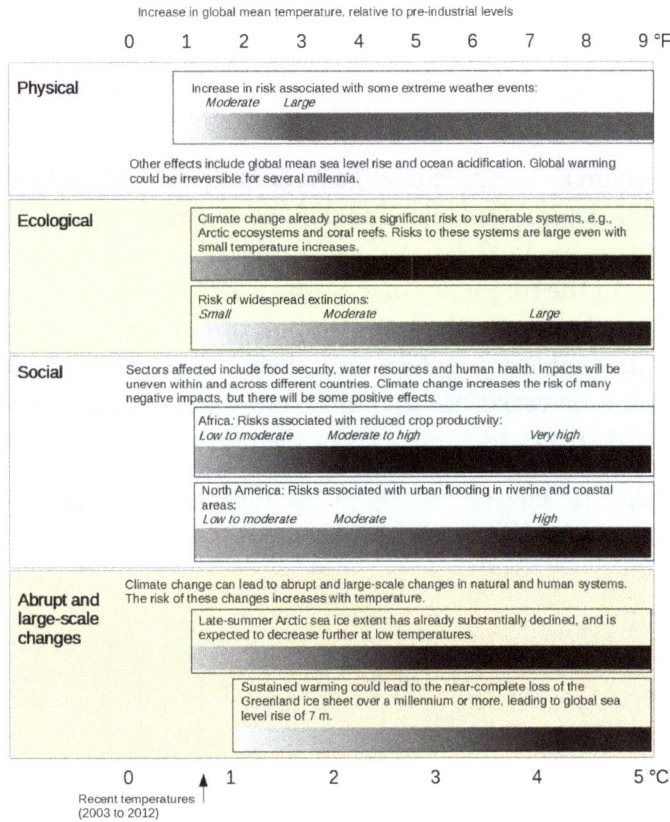

Summary of climate change impacts.

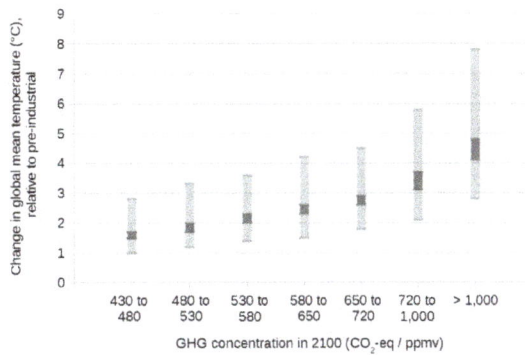

Projected global warming in 2100 for a range of emission scenarios.

Future effects of climate change will vary depending on climate change policies and social development. The two main policies to address climate change are reducing human greenhouse gas emissions (climate change mitigation) and adapting to the impacts of climate change. Geoengineering is another policy option.

Near-term climate change policies could significantly affect long-term climate change impacts. Stringent mitigation policies might be able to limit global warming (in 2100) to around 2 °C or below, relative to pre-industrial levels. Without mitigation, increased energy demand and extensive use of fossil fuels might lead to global warming of around 4 °C. Higher magnitudes of global warming would be more difficult to adapt to, and would increase the risk of negative impacts.

Definitions

In this article, "climate change" means a change in climate that persists over a sustained period of time. The World Meteorological Organization defines this time period as 30 years. Examples of climate change include increases in global surface temperature (global warming), changes in rainfall patterns, and changes in the frequency of extreme weather events. Changes in climate may be due to natural causes, e.g., changes in the sun's output, or due to human activities, e.g., changing the composition of the atmosphere. Any human-induced changes in climate will occur against a background of natural climatic variations and of variations in human activity such as population growth on shores or in arid areas which increase or decrease climate vulnerability.

Also, the term "anthropogenic forcing" refers to the influence exerted on a habitat or chemical environment by humans, as opposed to a natural process.

Temperature Changes

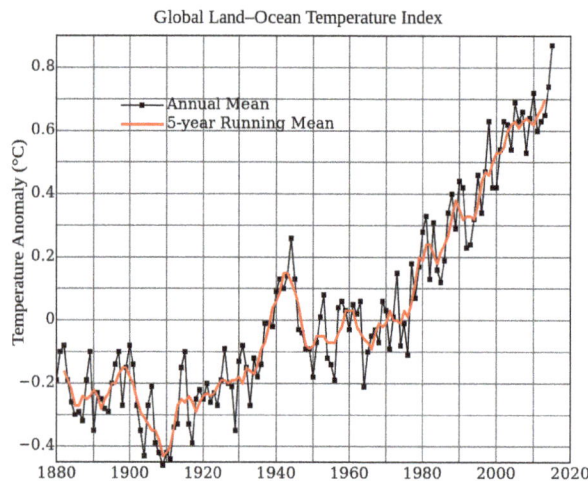

Global Land–Ocean Temperature Index

Global mean surface temperature change since 1880, relative to the 1951–1980 mean. Source: NASA GISS

The graph above shows the average of a set of temperature simulations for the 20th century (black line), followed by projected temperatures for the 21st century based on three greenhouse gas emissions scenarios (colored lines).

This article breaks down some of the impacts of climate change according to different levels of future global warming. This way of describing impacts has, for instance, been used in the IPCC (Intergovernmental Panel on Climate Change) Assessment Reports on climate change. The instrumental temperature record shows global warming of around 0.6 °C during the 20th century.

SRES Emissions Scenarios

The future level of global warming is uncertain, but a wide range of estimates (projections) have been made. The IPCC's "SRES" scenarios have been frequently used to make projections of future climate change. The SRES scenarios are "baseline" (or "reference") scenarios, which means that they do not take into account any current or future measures to limit GHG emissions (e.g., the UN-FCCC's Kyoto Protocol and the Cancún agreements). Emissions projections of the SRES scenarios are broadly comparable in range to the baseline emissions scenarios that have been developed by the scientific community.

In the IPCC Fourth Assessment Report, changes in future global mean temperature were projected using the six SRES "marker" emissions scenarios. Emissions projections for the six SRES "marker" scenarios are representative of the full set of forty SRES scenarios. For the lowest emissions SRES marker scenario, the best estimate for global mean temperature is an increase of 1.8 °C (3.2 °F) by the end of the 21st century. This projection is relative to global temperatures at the end of the 20th century. The "likely" range (greater than 66% probability, based on expert judgement) for the SRES B1 marker scenario is 1.1–2.9 °C (2.0–5.2 °F). For the highest emissions SRES marker scenario (A1FI), the best estimate for global mean temperature increase is 4.0 °C (7.2 °F), with a "likely" range of 2.4–6.4 °C (4.3–11.5 °F).

The range in temperature projections partly reflects (1) the choice of emissions scenario, and (2) the "climate sensitivity". For (1), different scenarios make different assumptions of future social and economic development (e.g., economic growth, population level, energy policies), which in turn affects projections of greenhouse gas (GHG) emissions. The projected magnitude of warming by 2100 is closely related to the level of cumulative emissions over the 21st century (i.e. total emissions between 2000-2100). The higher the cumulative emissions over this time period, the greater the level of warming is projected to occur.

(2) reflects uncertainty in the response of the climate system to past and future GHG emissions, which is measured by the climate sensitivity). Higher estimates of climate sensitivity lead to greater projected warming, while lower estimates of climate sensitivity lead to less projected warming.

Over the next several millennia, projections suggest that global warming could be irreversible. Even if emissions were drastically reduced, global temperatures would remain close to their highest level for at least 1,000 years.

Projected Warming in Context

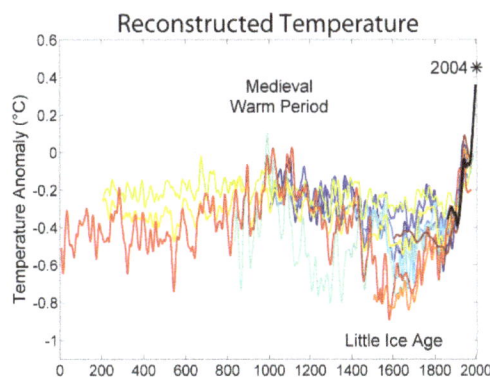

Two millennia of mean surface temperatures according to different reconstructions from climate proxies, each smoothed on a decadal scale, with the instrumental temperature record overlaid in black.

Global surface temperature for the past 5.3 million years as inferred from cores of ocean sediments taken all around the global ocean. The last 800,000 years are expanded in the lower half of the figure (image credit: NASA).

Scientists have used various "proxy" data to assess past changes in Earth's climate (paleoclimate). Sources of proxy data include historical records (such as farmers' logs), tree rings, corals, fossil pollen, ice cores, and ocean and lake sediments. Analysis of these data suggest that recent warming is unusual in the past 400 years, possibly longer. By the end of the 21st century, temperatures may increase to a level not experienced since the mid-Pliocene, around 3 million years ago. At that time, models suggest that mean global temperatures were about 2–3 °C warmer than pre-industrial temperatures. Even a 2 °C rise above the pre-industrial level would be outside the range of temperatures experienced by human civilization.

Physical Impacts

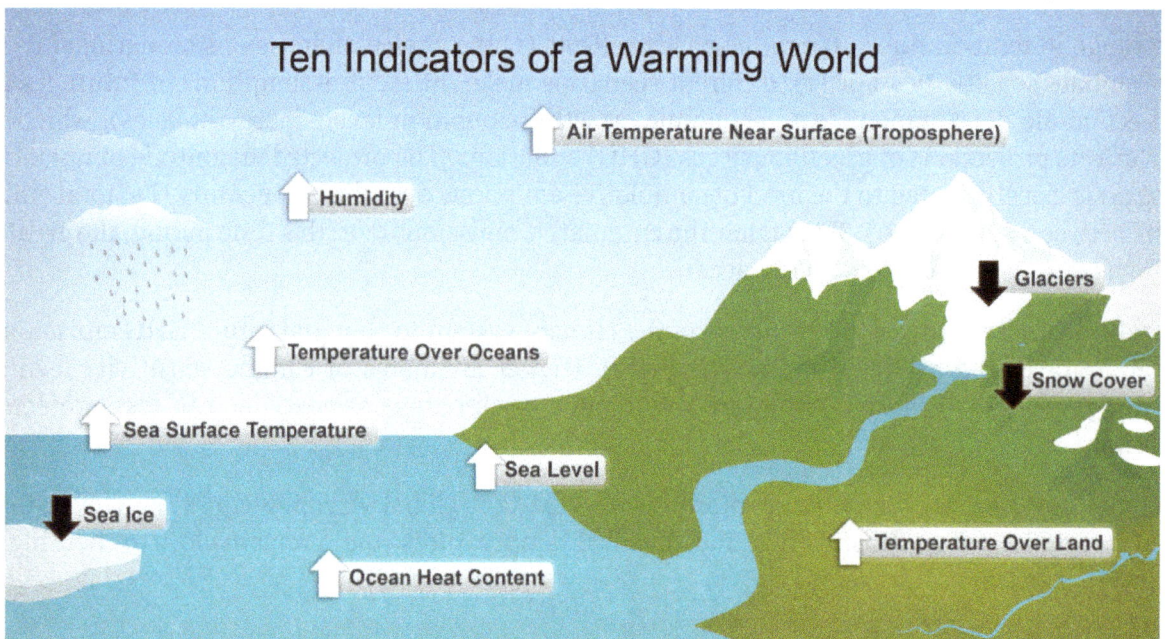

Ten Indicators of a Warming World

Seven of these indicators would be expected to increase in a warming world and observations show that they are, in fact, increasing. Three would be expected to decrease and they are, in fact, decreasing.

A broad range of evidence shows that the climate system has warmed. Evidence of global warming is shown in the graphs opposite. Some of the graphs show a positive trend, e.g., increasing temperature over land and the ocean, and sea level rise. Other graphs show a negative trend, e.g., decreased snow cover in the Northern Hemisphere, and declining Arctic sea ice extent. Evidence of warming is also apparent in living (biological) systems.

Human activities have contributed to a number of the observed changes in climate. This contribution has principally been through the burning of fossil fuels, which has led to an increase in the concentration of GHGs in the atmosphere. Another human influence on the climate are sulfur dioxide emissions, which are a precursor to the formation of sulfate aerosols in the atmosphere.

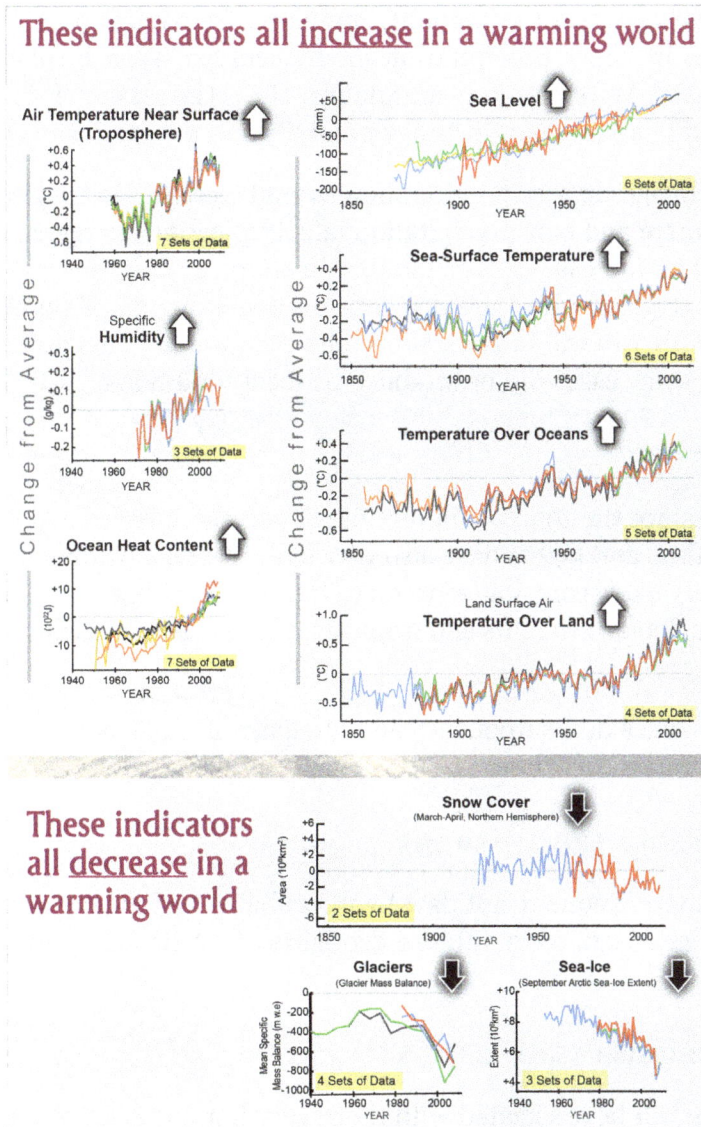

These indicators all increase in a warming world

Air Temperature Near Surface (Troposphere)

Sea Level

Specific Humidity

Sea-Surface Temperature

Ocean Heat Content

Temperature Over Oceans

Land Surface Air Temperature Over Land

These indicators all decrease in a warming world

Snow Cover (March-April, Northern Hemisphere)

Glaciers (Glacier Mass Balance)

Sea-Ice (September Arctic Sea-Ice Extent)

This set of graphs show changes in climate indicators over several decades. Each of the different colored lines in each panel represents an independently analyzed set of data. The data come from many different technologies including weather stations, satellites, weather balloons, ships and buoys.

Human-induced warming could lead to large-scale, irreversible, and/or abrupt changes in physical systems. An example of this is the melting of ice sheets, which contributes to sea level rise. The probability of warming having unforeseen consequences increases with the rate, magnitude, and duration of climate change.

Effects on Weather

Observations show that there have been changes in weather. As climate changes, the probabilities of certain types of weather events are affected.

Projected change in annual average precipitation by the end of the 21st century, based on a medium emissions scenario (SRES A1B) (Credit: NOAA Geophysical Fluid Dynamics Laboratory).

Changes have been observed in the amount, intensity, frequency, and type of precipitation. Widespread increases in heavy precipitation have occurred, even in places where total rain amounts have decreased. With medium confidence, IPCC (2012) concluded that human influences had contributed to an increase in heavy precipitation events at the global scale.

Projections of future changes in precipitation show overall increases in the global average, but with substantial shifts in where and how precipitation falls. Projections suggest a reduction in rainfall in the subtropics, and an increase in precipitation in subpolar latitudes and some equatorial regions. In other words, regions which are dry at present will in general become even drier, while regions that are currently wet will in general become even wetter. This projection does not apply to every locale, and in some cases can be modified by local conditions.

Extreme Weather

Over most land areas since the 1950s, it is very likely that there have been fewer or warmer cold days and nights. Hot days and nights have also very likely become warmer or more frequent. Human activities have very likely contributed to these trends. There may have been changes in other climate extremes (e.g., floods, droughts and tropical cyclones) but these changes are more difficult to identify.

Projections suggest changes in the frequency and intensity of some extreme weather events. Confidence in projections varies over time.

Near-term Projections (2016–2035)

Some changes (e.g., more frequent hot days) will probably be evident in the near term, while other near-term changes (e.g., more intense droughts and tropical cyclones) are more uncertain.

Long-term Projections (2081–2100)

Future climate change will be associated with more very hot days and fewer very cold days. The frequency, length and intensity of heat waves will very likely increase over most land areas. Higher growth in anthropogenic GHG emissions will be associated with larger increases in the frequency and severity of temperature extremes.

Assuming high growth in GHG emissions (IPCC scenario RCP8.5), presently dry regions may be affected by an increase in the risk of drought and reductions in soil moisture. Over most of the mid-latitude land masses and wet tropical regions, extreme precipitation events will very likely become more intense and frequent.

Tropical Cyclones

At the global scale, the frequency of tropical cyclones will probably decrease or be unchanged. Global mean tropical cyclone maximum wind speed and precipitation rates will likely increase. Changes in tropical cyclones will probably vary by region, but these variations are uncertain.

Effects of Climate Extremes

The impacts of extreme events on the environment and human society will vary. Some impacts will be beneficial—e.g., fewer cold extremes will probably lead to fewer cold deaths. Overall, however, impacts will probably be mostly negative.

Cryosphere

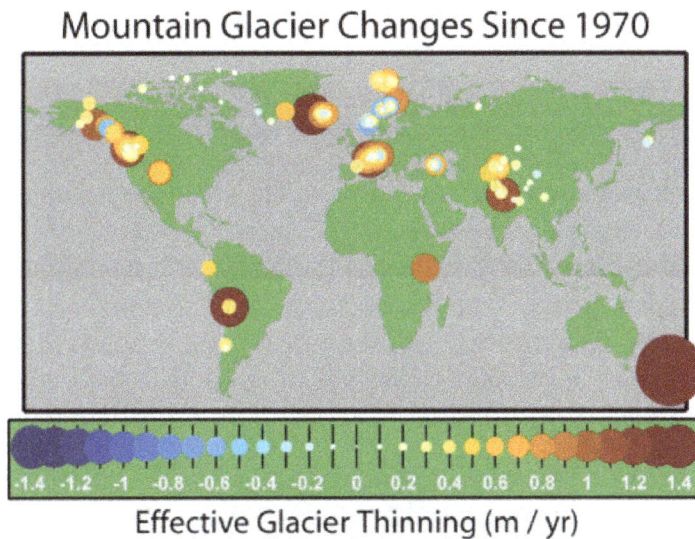

A map of the change in thickness of mountain glaciers since 1970. Thinning in orange and red, thickening in blue.

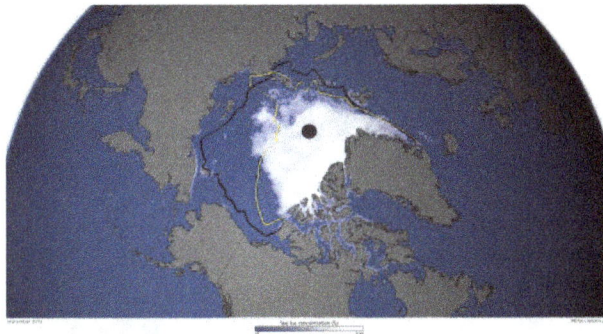

A map that shows ice concentration on 16 September 2012, along with the extent of the previous record low (yellow line) and the mid-September median extent (black line) setting a new record low that was 18 percent smaller than the previous record and nearly 50 percent smaller than the long-term (1979-2000) average.

The cryosphere is made up of areas of the Earth which are covered by snow or ice. Observed changes in the cryosphere include declines in Arctic sea ice extent, the widespread retreat of alpine glaciers, and reduced snow cover in the Northern Hemisphere.

Solomon *et al.* (2007) assessed the potential impacts of climate change on summertime Arctic sea ice extent. Assuming high growth in greenhouse gas emissions (SRES A2), some models projected that Arctic sea ice in the summer could largely disappear by the end of the 21st century. More recent projections suggest that the Arctic summers could be ice-free (defined as ice extent less than 1 million square km) as early as 2025-2030.

During the 21st century, glaciers and snow cover are projected to continue their widespread retreat. In the western mountains of North America, increasing temperatures and changes in precipitation are projected to lead to reduced snowpack. Snowpack is the seasonal accumulation of slow-melting snow. The melting of the Greenland and West Antarctic ice sheets could contribute to sea level rise, especially over long time-scales.

Changes in the cryosphere are projected to have social impacts. For example, in some regions, glacier retreat could increase the risk of reductions in seasonal water availability. Barnett *et al.* (2005) estimated that more than one-sixth of the world's population rely on glaciers and snowpack for their water supply.

Oceans

The role of the oceans in global warming is complex. The oceans serve as a sink for carbon dioxide, taking up much that would otherwise remain in the atmosphere, but increased levels of CO2 have led to ocean acidification. Furthermore, as the temperature of the oceans increases, they become less able to absorb excess CO2. The ocean have also acted as a sink in absorbing extra heat from the atmosphere. The increase in ocean heat content is much larger than any other store of energy in the Earth's heat balance over the two periods 1961 to 2003 and 1993 to 2003, and accounts for more than 90% of the possible increase in heat content of the Earth system during these periods.

Global warming is projected to have a number of effects on the oceans. Ongoing effects include rising sea levels due to thermal expansion and melting of glaciers and ice sheets, and warming of the ocean surface, leading to increased temperature stratification. Other possible effects include large-scale changes in ocean circulation.

Acidification

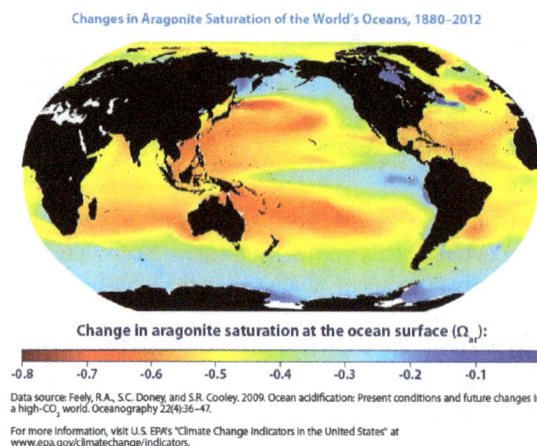

Changes in Aragonite Saturation of the World's Oceans, 1880–2012

Change in aragonite saturation at the ocean surface (Ω_{ar}):

| -0.8 | -0.7 | -0.6 | -0.5 | -0.4 | -0.3 | -0.2 | -0.1 | 0 |

Data source: Feely, R.A., S.C. Doney, and S.R. Cooley. 2009. Ocean acidification: Present conditions and future changes in a high-CO$_2$ world. Oceanography 22(4):36–47.

For more information, visit U.S. EPA's "Climate Change Indicators in the United States" at www.epa.gov/climatechange/indicators.

This map shows changes in the amount of aragonite dissolved in ocean surface waters between the 1880s and the most recent decade (2003-2012). Historical modeling suggests that since the 1880s, increased CO$_2$ has led to lower aragonite saturation levels (less availability of minerals) in the oceans around the world. The largest decreases in aragonite saturation have occurred in tropical waters. However, decreases in cold areas may be of greater concern because colder waters typically have lower aragonite levels to begin with.

About one-third of the carbon dioxide emitted by human activity has already been taken up by the oceans. As carbon dioxide dissolves in sea water, carbonic acid is formed, which has the effect of acidifying the ocean, measured as a change in pH. The uptake of human carbon emissions since the year 1750 has led to an average decrease in pH of 0.1 units. Projections using the SRES emissions scenarios suggest a further reduction in average global surface ocean pH of between 0.14 and 0.35 units over the 21st century.

The effects of ocean acidification on the marine biosphere have yet to be documented. Laboratory experiments suggest beneficial effects for a few species, with potentially highly detrimental effects for a substantial number of species. With medium confidence, Fischlin *et al.* (2007) projected that future ocean acidification and climate change would impair a wide range of planktonic and shallow benthic marine organisms that use aragonite to make their shells or skeletons, such as corals and marine snails (pteropods), with significant impacts particularly in the Southern Ocean.

Oxygen Depletion

The amount of oxygen dissolved in the oceans may decline, with adverse consequences for ocean life.

Sea Level Rise

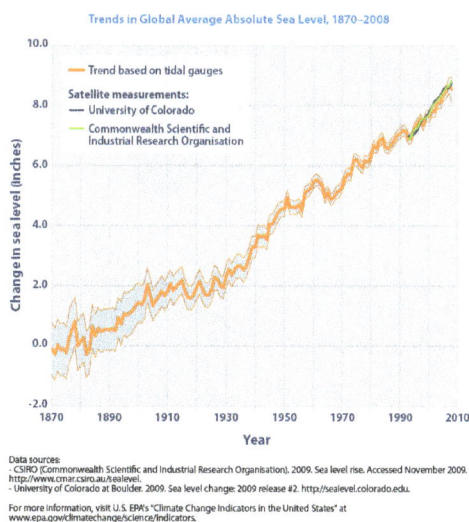

Trends in global average absolute sea level, 1870-2008.

There is strong evidence that global sea level rose gradually over the 20th century. With high confidence, Bindoff *et al.* (2007) concluded that between the mid-19th and mid-20th centuries, the rate of sea level rise increased. Authors of the IPCC Fourth Assessment Synthesis Report (IPCC AR4 SYR, 2007) reported that between the years 1961 and 2003, global average sea level rose at an average rate of 1.8 mm per year (mm/yr), with a range of 1.3–2.3 mm/yr. Between 1993 and 2003, the rate increased above the previous period to 3.1 mm/yr (range of 2.4–3.8 mm/yr). Authors of IPCC AR4 SYR (2007) were uncertain whether the increase in rate from 1993 to 2003 was due to natural variations in sea level over the time period, or whether it reflected an increase in the underlying long-term trend.

There are two main factors that have contributed to observed sea level rise. The first is thermal expansion: as ocean water warms, it expands. The second is from the contribution of land-based ice due to increased melting. The major store of water on land is found in glaciers and ice sheets. Anthropogenic forces very likely (greater than 90% probability, based on expert judgement) contributed to sea level rise during the latter half of the 20th century.

There is a widespread consensus that substantial long-term sea level rise will continue for centuries to come. In their Fourth Assessment Report, the IPCC projected sea level rise to the end of the 21st century using the SRES emissions scenarios. Across the six SRES marker scenarios, sea level was projected to rise by 18 to 59 cm (7.1 to 23.2 in), relative to sea level at the end of the 20th century. Thermal expansion is the largest component in these projections, contributing 70-75% of the central estimate for all scenarios. Due to a lack of scientific understanding, this sea level rise estimate does not include all of the possible contributions of ice sheets.

An assessment of the scientific literature on climate change was published in 2010 by the US National Research Council (US NRC, 2010). NRC (2010) described the projections in AR4 (i.e. those cited in the above paragraph) as "conservative", and summarized the results of more recent studies. Cited studies suggested a great deal of uncertainty in projections. A range of projections suggested possible sea level rise by the end of the 21st century of between 0.56 and 2 m, relative to sea levels at the end of the 20th century.

Ocean Temperature Rise

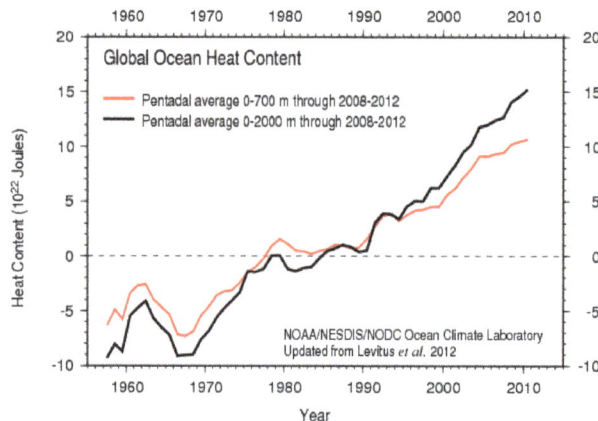

Global ocean heat content from 1955-2012

From 1961 to 2003, the global ocean temperature has risen by 0.10 °C from the surface to a depth of 700 m. There is variability both year-to-year and over longer time scales, with global ocean heat content observations showing high rates of warming for 1991–2003, but some cooling from 2003 to 2007. The temperature of the Antarctic Southern Ocean rose by 0.17 °C (0.31 °F) between the 1950s and the 1980s, nearly twice the rate for the world's oceans as a whole. As well as having effects on ecosystems (e.g. by melting sea ice, affecting algae that grow on its underside), warming reduces the ocean's ability to absorb CO_2. It is likely (greater than 66% probability, based on expert judgement) that anthropogenic forcing contributed to the general warming observed in the upper several hundred metres of the ocean during the latter half of the 20th century.

Regions

Temperatures across the world in the 1880s (left) and the 1980s (right), as compared to average temperatures from 1951 to 1980.

Projected changes in average temperatures across the world in the 2050s under three greenhouse gas (GHG) emissions scenarios.

Regional effects of global warming vary in nature. Some are the result of a generalised global change, such as rising temperature, resulting in local effects, such as melting ice. In other cases, a change may be related to a change in a particular ocean current or weather system. In such cases, the regional effect may be disproportionate and will not necessarily follow the global trend.

There are three major ways in which global warming will make changes to regional climate: melting or forming ice, changing the hydrological cycle (of evaporation and precipitation) and changing currents in the oceans and air flows in the atmosphere. The coast can also be considered a region, and will suffer severe impacts from sea level rise.

Observed Impacts

With very high confidence, Rosenzweig *et al.* (2007) concluded that physical and biological systems on all continents and in most oceans had been affected by recent climate changes, particularly regional temperature increases. Impacts include earlier leafing of trees and plants over many regions; movements of species to higher latitudes and altitudes in the Northern Hemisphere; changes in bird migrations in Europe, North America and Australia; and shifting of the oceans' plankton and fish from cold- to warm-adapted communities.

The human influence on the climate can be seen in the geographical pattern of observed warming, with greater temperature increases over land and in polar regions rather than over the oceans. Using models, it is possible to identify the human "signal" of global warming over both land and ocean areas.

Projected Impacts

Projections of future climate changes at the regional scale do not hold as high a level of scientific confidence as projections made at the global scale. It is, however, expected that future warming will follow a similar geographical pattern to that seen already, with greatest warming over land and high northern latitudes, and least over the Southern Ocean and parts of the North Atlantic Ocean. Nearly all land areas will very likely warm more than the global average.

The Arctic, Africa, small islands and Asian megadeltas are regions that are likely to be especially affected by climate change. Low-latitude, less-developed areas are at most risk of experiencing negative impacts due to climate change. Developed countries are also vulnerable to climate change. For example, developed countries will be negatively affected by increases in the severity and frequency of some extreme weather events, such as heat waves. In all regions, some people can be particularly at risk from climate change, such as the poor, young children and the elderly.

Social Systems

The impacts of climate change can be thought of in terms of sensitivity and vulnerability. "Sensitivity" is the degree to which a particular system or sector might be affected, positively or negatively, by climate change and/or climate variability. "Vulnerability" is the degree to which a particular system or sector might be adversely affected by climate change.

The sensitivity of human society to climate change varies. Sectors sensitive to climate change include water resources, coastal zones, human settlements, and human health. Industries sensitive to climate change include agriculture, fisheries, forestry, energy, construction, insurance, financial services, tourism, and recreation.

Food Supply

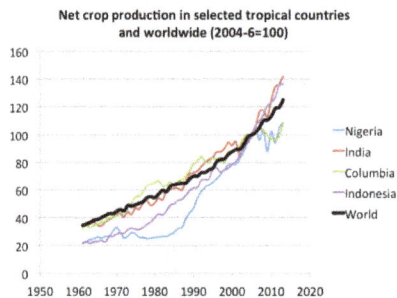

Graph of net crop production worldwide and in selected tropical countries. Raw data from the United Nations.

Climate change will impact agriculture and food production around the world due to: the effects of elevated CO_2 in the atmosphere, higher temperatures, altered precipitation and transpiration regimes, increased frequency of extreme events, and modified weed, pest, and pathogen pressure. In general, low-latitude areas are at most risk of having decreased crop yields.

As of 2007, the effects of regional climate change on agriculture have been small. Changes in crop phenology provide important evidence of the response to recent regional climate change. Phenology is the study of natural phenomena that recur periodically, and how these phenomena relate to climate and seasonal changes. A significant advance in phenology has been observed for agriculture and forestry in large parts of the Northern Hemisphere.

Projections

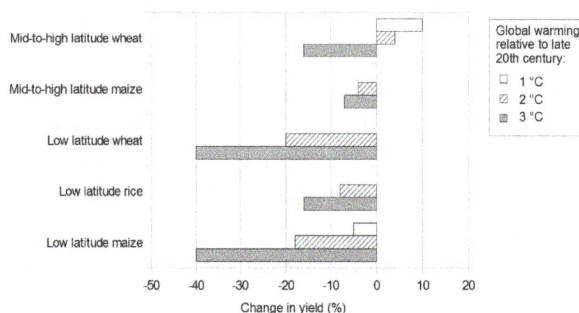

Projected changes in crop yields at different latitudes with global warming. This graph is based on several studies.

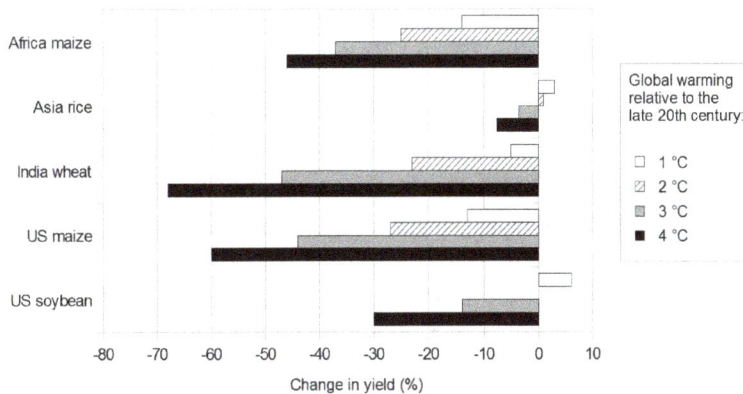

Projected changes in yields of selected crops with global warming. This graph is based on several studies.

With low to medium confidence, Schneider *et al.* (2007) projected that for about a 1 to 3 °C increase in global mean temperature (by the years 2090-2100, relative to average temperatures in the years 1990–2000), there would be productivity decreases for some cereals in low latitudes, and productivity increases in high latitudes. With medium confidence, global production potential was projected to:

- increase up to around 3 °C,

- very likely decrease above about 3 °C.

Most of the studies on global agriculture assessed by Schneider *et al.* (2007) had not incorporated a number of critical factors, including changes in extreme events, or the spread of pests and diseases. Studies had also not considered the development of specific practices or technologies to aid adaptation to climate change.

The graphs opposite show the projected effects of climate change on selected crop yields. Actual changes in yields may be above or below these central estimates.

The projections above can be expressed relative to pre-industrial (1750) temperatures. 0.6 °C of warming is estimated to have occurred between 1750 and 1990-2000. Add 0.6 °C to the above projections to convert them from a 1990-2000 to pre-industrial baseline.

Food Security

Easterling *et al.* (2007) assessed studies that made quantitative projections of climate change impacts on food security. It was noted that these projections were highly uncertain and had limitations. However, the assessed studies suggested a number of fairly robust findings. The first was that climate change would likely increase the number of people at risk of hunger compared with reference scenarios with no climate change. Climate change impacts depended strongly on projected future social and economic development. Additionally, the magnitude of climate change impacts was projected to be smaller compared to the impact of social and economic development. In 2006, the global estimate for the number of people undernourished was 820 million. Under the SRES A1, B1, and B2 scenarios, projections for the year 2080 showed a reduction in the number of people undernourished of about 560-700 million people, with a global total of undernourished people of 100-240 million in 2080. By contrast, the SRES A2 scenario showed only a small decrease in the risk of hunger from 2006 levels. The smaller reduction under A2 was attributed to the higher projected future population level in this scenario.

Droughts and Agriculture

Some evidence suggests that droughts have been occurring more frequently because of global warming and they are expected to become more frequent and intense in Africa, southern Europe, the Middle East, most of the Americas, Australia, and Southeast Asia. However, other research suggests that there has been little change in drought over the past 60 years. Their impacts are aggravated because of increased water demand, population growth, urban expansion, and environmental protection efforts in many areas. Droughts result in crop failures and the loss of pasture grazing land for livestock.

Health

Human beings are exposed to climate change through changing weather patterns (temperature, precipitation, sea-level rise and more frequent extreme events) and indirectly through changes in water, air and food quality and changes in ecosystems, agriculture, industry and settlements and the economy (Confalonieri *et al.*, 2007:393). According to an assessment of the scientific literature by Confalonieri *et al.* (2007:393), the effects of climate change to date have been small, but are projected to progressively increase in all countries and regions.

A study by the World Health Organization (WHO, 2009) estimated the effect of climate change on human health. Not all of the effects of climate change were included in their estimates, for example, the effects of more frequent and extreme storms were excluded. Climate change was estimated to have been responsible for 3% of diarrhoea, 3% of malaria, and 3.8% of dengue fever deaths worldwide in 2004. Total attributable mortality was about 0.2% of deaths in 2004; of these, 85% were child deaths.

Projections

With high confidence, authors of the IPCC AR4 Synthesis report projected that climate change would bring some benefits in temperate areas, such as fewer deaths from cold exposure, and some mixed effects such as changes in range and transmission potential of malaria in Africa. Benefits were projected to be outweighed by negative health effects of rising temperatures, especially in developing countries.

With very high confidence, Confalonieri *et al.* (2007) concluded that economic development was an important component of possible adaptation to climate change. Economic growth on its own, however, was not judged to be sufficient to insulate the world's population from disease and injury due to climate change. Future vulnerability to climate change will depend not only on the extent of social and economic change, but also on how the benefits and costs of change are distributed in society. For example, in the 19th century, rapid urbanization in western Europe lead to a plummeting in population health. Other factors important in determining the health of populations include education, the availability of health services, and public-health infrastructure.

Water Resources

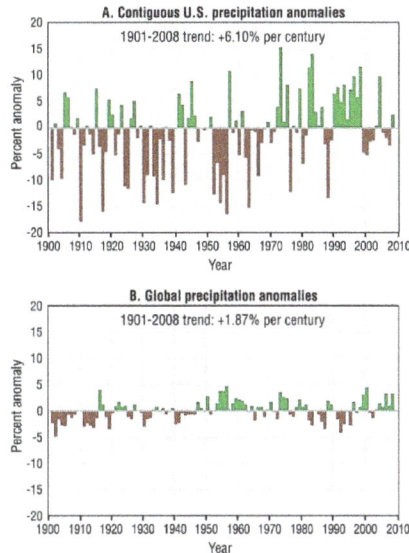

ªAnomalies and percent change are calculated with respect to the
1971-2000 mean.
Data source: NOAA, 2009

Precipitation during the 20th century and up through 2008 during global warming, the NOAA estimating an observed trend over that period of 1.87% global precipitation increase per century.

A number of climate-related trends have been observed that affect water resources. These include changes in precipitation, the crysosphere and surface waters (e.g., changes in river flows). Observed and projected impacts of climate change on freshwater systems and their management are mainly due to changes in temperature, sea level and precipitation variability. Sea level rise will extend areas of salinization of groundwater and estuaries, resulting in a decrease in freshwater availability for humans and ecosystems in coastal areas. In an assessment of the scientific literature, Kundzewicz *et al.* (2007) concluded, with high confidence, that:

the negative impacts of climate change on freshwater systems outweigh the benefits. All of the regions assessed in the IPCC Fourth Assessment Report (Africa, Asia, Australia and New Zealand, Europe, Latin America, North America, Polar regions (Arctic and Antarctic), and small islands) showed an overall net negative impact of climate change on water resources and freshwater ecosystems.

Semi-arid and arid areas are particularly exposed to the impacts of climate change on freshwater. With very high confidence, it was judged that many of these areas, e.g., the Mediterranean basin, Western United States, Southern Africa, and north-eastern Brazil, would suffer a decrease in water resources due to climate change.

Migration and Conflict

General circulation models project that the future climate change will bring wetter coasts, drier mid-continent areas, and further sea level rise. Such changes could result in the gravest effects of climate change through human migration. Millions might be displaced by shoreline erosions, river and coastal flooding, or severe drought.

Migration related to climate change is likely to be predominantly from rural areas in developing countries to towns and cities. In the short term climate stress is likely to add incrementally to existing migration patterns rather than generating entirely new flows of people.

It has been argued that environmental degradation, loss of access to resources (e.g., water resources), and resulting human migration could become a source of political and even military conflict. Factors other than climate change may, however, be more important in affecting conflict. For example, Wilbanks *et al.* (2007) suggested that major environmentally influenced conflicts in Africa were more to do with the relative abundance of resources, e.g., oil and diamonds, than with resource scarcity. Scott *et al.* (2001) placed only low confidence in predictions of increased conflict due to climate change.

A 2013 study found that significant climatic changes were associated with a higher risk of conflict worldwide, and predicted that "amplified rates of human conflict could represent a large and critical social impact of anthropogenic climate change in both low- and high-income countries." Similarly, a 2014 study found that higher temperatures were associated with a greater likelihood of violent crime, and predicted that global warming would cause millions of such crimes in the United States alone during the 21st century.

Military planners are concerned that global warming is a "threat multiplier". "Whether it is poverty, food and water scarcity, diseases, economic instability, or threat of natural disasters, the broad range of changing climatic conditions may be far reaching. These challenges may threaten stability in much of the world".

Aggregate Impacts

Aggregating impacts adds up the total impact of climate change across sectors and/or regions. Examples of aggregate measures include economic cost (e.g., changes in gross domestic product (GDP) and the social cost of carbon), changes in ecosystems (e.g., changes over land area from one type of vegetation to another), human health impacts, and the number of people affected by climate change. Aggregate measures such as economic cost require researchers to make value judgements over the importance of impacts occurring in different regions and at different times.

Observed Impacts

Global losses reveal rapidly rising costs due to extreme weather-related events since the 1970s. Socio-economic factors have contributed to the observed trend of global losses, e.g., population growth, increased wealth. Part of the growth is also related to regional climatic factors, e.g., changes in precipitation and flooding events. It is difficult to quantify the relative impact of socio-economic factors and climate change on the observed trend. The trend does, however, suggest increasing vulnerability of social systems to climate change.

Projected Impacts

The total economic impacts from climate change are highly uncertain. With medium confidence, Smith *et al.* (2001) concluded that world GDP would change by plus or minus a few percent for a small increase in global mean temperature (up to around 2 °C relative to the 1990 temperature level). Most studies assessed by Smith *et al.* (2001) projected losses in world GDP for a medium increase in global mean temperature (above 2-3 °C relative to the 1990 temperature level), with increasing losses for greater temperature increases. This assessment is consistent with the findings of more recent studies, as reviewed by Hitz and Smith (2004).

Economic impacts are expected to vary regionally. For a medium increase in global mean temperature (2-3 °C of warming, relative to the average temperature between 1990–2000), market sectors in low-latitude and less-developed areas might experience net costs due to climate change. On the other hand, market sectors in high-latitude and developed regions might experience net benefits for this level of warming. A global mean temperature increase above about 2-3 °C (relative to 1990-2000) would very likely result in market sectors across all regions experiencing either declines in net benefits or rises in net costs.

Aggregate impacts have also been quantified in non-economic terms. For example, climate change over the 21st century is likely to adversely affect hundreds of millions of people through increased coastal flooding, reductions in water supplies, increased malnutrition and increased health impacts.

Biological Systems

Observed Impacts on Biological Systems

A vast array of physical and biological systems across the Earth are being affected by human-induced global warming.

With very high confidence, Rosenzweig *et al.* (2007) concluded that recent warming had strongly affected natural biological systems. Hundreds of studies have documented responses of ecosystems, plants, and animals to the climate changes that have already occurred. For example, in the Northern Hemisphere, species are almost uniformly moving their ranges northward and up in elevation in search of cooler temperatures. Humans are very likely causing changes in regional temperatures to which plants and animals are responding.

Projected Impacts on Biological Systems

By the year 2100, ecosystems will be exposed to atmospheric CO_2 levels substantially higher than in the past 650,000 years, and global temperatures at least among the highest of those experienced in the past 740,000 years. Significant disruptions of ecosystems are projected to increase with future climate change. Examples of disruptions include disturbances such as fire, drought, pest infestation, invasion of species, storms, and coral bleaching events. The stresses caused by climate change, added to other stresses on ecological systems (e.g., land conversion, land degradation, harvesting, and pollution), threaten substantial damage to or complete loss of some unique ecosystems, and extinction of some critically endangered species.

Climate change has been estimated to be a major driver of biodiversity loss in cool conifer forests, savannas, mediterranean-climate systems, tropical forests, in the Arctic tundra, and in coral reefs. In other ecosystems, land-use change may be a stronger driver of biodiversity loss at least in the near-term. Beyond the year 2050, climate change may be the major driver for biodiversity loss globally.

A literature assessment by Fischlin *et al.* (2007) included a quantitative estimate of the number of species at increased risk of extinction due to climate change. With medium confidence, it was projected that approximately 20 to 30% of plant and animal species assessed so far (in an unbiased sample) would likely be at increasingly high risk of extinction should global mean temperatures exceed a warming of 2 to 3 °C above pre-industrial temperature levels. The uncertainties in this estimate, however, are large: for a rise of about 2 °C the percentage may be as low as 10%, or for about 3 °C, as high as 40%, and depending on biota (all living organisms of an area, the flora and fauna considered as a unit) the range is between 1% and 80%. As global average temperature exceeds 4 °C above pre-industrial levels, model projections suggested that there could be significant extinctions (40-70% of species that were assessed) around the globe.

Assessing whether future changes in ecosystems will be beneficial or detrimental is largely based on how ecosystems are valued by human society. For increases in global average temperature exceeding 1.5 to 2.5 °C (relative to global temperatures over the years 1980-1999) and in concomitant atmospheric CO_2 concentrations, projected changes in ecosystems will have predominantly negative consequences for biodiversity and ecosystems goods and services, e.g., water and food supply.

Abrupt or Irreversible Changes

Physical, ecological and social systems may respond in an abrupt, non-linear or irregular way to climate change. This is as opposed to a smooth or regular response. A quantitative entity behaves "irregularly" when its dynamics are discontinuous (i.e., not smooth), nondifferentiable, unbounded, wildly varying, or otherwise ill-defined. Such behaviour is often termed "singular". Irregular behaviour in Earth systems may give rise to certain thresholds, which, when crossed, may lead to a large change in the system.

Some singularities could potentially lead to severe impacts at regional or global scales. Examples of "large-scale" singularities are discussed in the articles on abrupt climate change, climate change feedback and runaway climate change. It is possible that human-induced climate change could trigger large-scale singularities, but the probabilities of triggering such events are, for the most part, poorly understood.

With low to medium confidence, Smith *et al.* (2001) concluded that a rapid warming of more than 3 °C above 1990 levels would exceed thresholds that would lead to large-scale discontinuities in the climate system. Since the assessment by Smith *et al.* (2001), improved scientific understanding provides more guidance for two large-scale singularities: the role of carbon cycle feedbacks in future climate change (discussed below in the section on biogeochemical cycles) and the melting of the Greenland and West Antarctic ice sheets.

Biogeochemical Cycles

Climate change may have an effect on the carbon cycle in an interactive "feedback" process. A feedback exists where an initial process triggers changes in a second process that in turn influences the initial process. A positive feedback intensifies the original process, and a negative feedback reduces it. Models suggest that the interaction of the climate system and the carbon cycle is one where the feedback effect is positive.

Using the A2 SRES emissions scenario, Schneider *et al.* (2007) found that this effect led to additional warming by the years 2090-2100 (relative to the 1990–2000) of 0.1–1.5 °C. This estimate was made with high confidence. The climate projections made in the IPCC Fourth Assessment Report summarized earlier of 1.1–6.4 °C account for this feedback effect. On the other hand, with medium confidence, Schneider *et al.* (2007) commented that additional releases of GHGs were possible from permafrost, peat lands, wetlands, and large stores of marine hydrates at high latitudes.

Greenland and West Antarctic Ice sheets

With medium confidence, authors of AR4 concluded that with a global average temperature increase of 1–4 °C (relative to temperatures over the years 1990–2000), at least a partial deglaciation of the Greenland ice sheet, and possibly the West Antarctic ice sheets would occur. The estimated timescale for partial deglaciation was centuries to millennia, and would contribute 4 to 6 metres (13 to 20 ft) or more to sea level rise over this period.

Atlantic Meridional Overturning Circulation

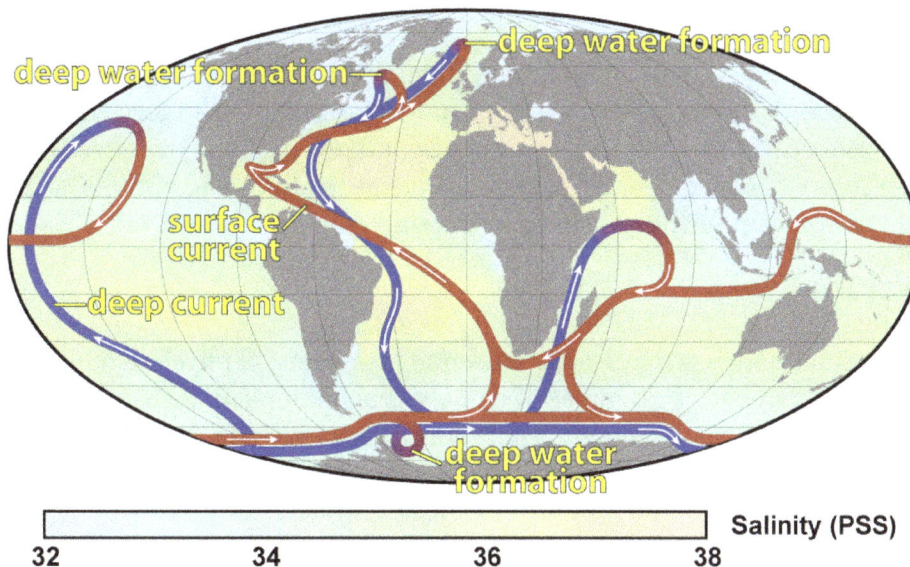

This map shows the general location and direction of the warm surface (red) and cold deep water (blue) currents of the thermohaline circulation. Salinity is represented by color in units of the Practical Salinity Scale. Low values (blue) are less saline, while high values (orange) are more saline.

The Atlantic Meridional Overturning Circulation (AMOC) is an important component of the Earth's climate system, characterized by a northward flow of warm, salty water in the upper layers of the Atlantic and a southward flow of colder water in the deep Atlantic. The AMOC is equivalently known as the thermohaline circulation (THC). Potential impacts associated with MOC changes include reduced warming or (in the case of abrupt change) absolute cooling of northern high-latitude areas near Greenland and north-western Europe, an increased warming of Southern Hemisphere high-latitudes, tropical drying, as well as changes to marine ecosystems, terrestrial vegetation, oceanic

CO2 uptake, oceanic oxygen concentrations, and shifts in fisheries. According to an assessment by the US Climate Change Science Program (CCSP, 2008b), it is very likely (greater than 90% probability, based on expert judgement) that the strength of the AMOC will decrease over the course of the 21st century. Warming is still expected to occur over most of the European region downstream of the North Atlantic Current in response to increasing GHGs, as well as over North America. Although it is very unlikely (less than 10% probability, based on expert judgement) that the AMOC will collapse in the 21st century, the potential consequences of such a collapse could be severe.

Irreversibilities

Commitment to Radiative Forcing

Emissions of GHGs are a potentially irreversible commitment to sustained radiative forcing in the future. The contribution of a GHG to radiative forcing depends on the gas's ability to trap infrared (heat) radiation, the concentration of the gas in the atmosphere, and the length of time the gas resides in the atmosphere.

CO2 is the most important anthropogenic GHG. While more than half of the CO2 emitted is currently removed from the atmosphere within a century, some fraction (about 20%) of emitted CO2 remains in the atmosphere for many thousands of years. Consequently, CO2 emitted today is potentially an irreversible commitment to sustained radiative forcing over thousands of years.

This commitment may not be truly irreversible should techniques be developed to remove CO2 or other GHGs directly from the atmosphere, or to block sunlight to induce cooling. Techniques of this sort are referred to as geoengineering. Little is known about the effectiveness, costs or potential side-effects of geoengineering options. Some geoengineering options, such as blocking sunlight, would not prevent further ocean acidification.

Irreversible Impacts

Human-induced climate change may lead to irreversible impacts on physical, biological, and social systems. There are a number of examples of climate change impacts that may be irreversible, at least over the timescale of many human generations. These include the large-scale singularities described above – changes in carbon cycle feedbacks, the melting of the Greenland and West Antarctic ice sheets, and changes to the AMOC. In biological systems, the extinction of species would be an irreversible impact. In social systems, unique cultures may be lost due to climate change. For example, humans living on atoll islands face risks due to sea-level rise, sea-surface warming, and increased frequency and intensity of extreme weather events.

Benefits of Global Warming

With a large range of effects, it is unlikely that all effects will be negative. Not only have some positive effects been identified, but there is some published material indicating that a small amount of warming would be good. The IPCC cautions that "Estimates agree on the size of the impact (small relative to economic growth), and 17 of the 20 impact estimates shown in Figure 10-1 are negative. Losses accelerate with greater warming, and estimates diverge."

The identified benefits are listed below.

CO2 Fertilisation Effect

CO2 is one of the substances which plants require to grow. Increasing its amount in the air contributes to:

- Improved agriculture in some high latitude regions
- Increased growing season in Greenland
- Increased productivity of sour orange trees
- Increased vegetation activity in high northern latitudes
- Increased plankton biomass in the North Pacific Subtropical Gyre
- Recent increase in forest growth
- Increased Arctic tundra plant reproduction

Human Health

Winter deaths will decline as temperatures warm

Ice-free Northwest Passage

Ships will travel on a shorter route between the Pacific and Atlantic oceans

Animal Population Changes

- Some animals will benefit from the warming:
- Increase in chinstrap and gentoo penguins
- Bigger marmots

Scientific Opinion

The Intergovernmental Panel on Climate Change (IPCC) has published several major assessments on the effects of global warming. Its most recent comprehensive impact assessment was published in 2014. Publications describing the effects of climate change have also been produced by the following organizations:

- American Association for the Advancement of Science (AAAS)
- A report by the Netherlands Environmental Assessment Agency, the Royal Netherlands Meteorological Institute, and Wageningen University and Research Centre
- UK AVOID research programme
- A report by the UK Royal Society and US National Academy of Sciences
- University of New South Wales Climate Change Research Centre
- US National Research Council

A report by Molina *et al.* (no date) states:

The overwhelming evidence of human-caused climate change documents both current impacts with significant costs and extraordinary future risks to society and natural systems

NASA Data and Tools

NASA has released public data and tools to predict how temperature and rainfall patterns worldwide may change through to the year 2100 caused by increasing carbon dioxide in Earth's atmosphere. The dataset shows projected changes worldwide on a regional level simulated by 21 climate models. The data can be viewed on a daily timescale for individual cities and towns and may be used to conduct climate risk assessments to predict the local and global effects of weather dangers, for example droughts, floods, heat waves and declines in agriculture productivity, and help plan responses to global warming effects.

References

- IPCC SAR WG2 (1996), Watson, R.T.; Zinyowera, M.C.; Moss, R.H., eds., Climate Change 1995: Impacts, Adaptations and Mitigation of Climate Change: Scientific-Technical Analyses, Contribution of Working Group II to the Second Assessment Report of the Intergovernmental Panel on Climate Change, Cambridge University Press, ISBN 0-521-56431-X.

- Committee on Ecological Impacts of Climate Change, US National Research Council (NRC) (2008). Ecological Impacts of Climate Change. 500 Fifth Street, NW Washington, DC 20001, USA: The National Academies Press. ISBN 978-0-309-12710-3.

- IPCC TAR WG1 (2001), Houghton, J.T.; Ding, Y.; Griggs, D.J.; Noguer, M.; van der Linden, P.J.; Dai, X.; Maskell, K.; Johnson, C.A., eds., Climate Change 2001: The Scientific Basis, Contribution of Working Group I to the Third Assessment Report of the Intergovernmental Panel on Climate Change, Cambridge University Press, ISBN 0-521-80767-0.

- Karl, Thomas R.; Melillo, Jerry M.; Peterson, Thomas C., eds. (2009). Global Climate Change Impacts in the United States (PDF). New York: Cambridge University Press. ISBN 978-0-521-14407-0.

- US NRC (2010). Advancing the Science of Climate Change. A report by the US National Research Council (NRC). Washington, D.C., USA: National Academies Press. ISBN 0-309-14588-0.

- Maslowski, Wieslaw (16 March 2010). "Advancements and Limitations in Understanding and Predicting Arctic Climate Change". State of the Arctic (conference website). Retrieved 2 February 2015.

- Jin Liao et al.(2013) (January 2014). "High levels of molecular chlorine in the Arctic atmosphere". Nature Geoscience Letter. Bibcode:2014NatGe...7...91L. doi:10.1038/ngeo2046. Retrieved January 14, 2014.

- J A Screen (November 2013). "Influence of Arctic sea ice on European summer precipitation". Environmental Research Letters 8 (4). Bibcode:2013ERL.....8d4015S. doi:10.1088/1748-9326/8/4/044015. Retrieved January 26, 2014.

- Qiuhong Tang; Xuejun Zhang; Jennifer A. Francis (December 2013). "Extreme summer weather in northern mid-latitudes linked to a vanishing cryosphere". Nature Climate Change 4: 45–50. Bibcode:2014NatCC...4...45T. doi:10.1038/nclimate2065. Retrieved January 26, 2014.

- James E. Overland (December 2013). "Atmospheric science: Long-range linkage". Nature Climate Change 4 (1): 11–12. Bibcode:2014NatCC...4...11O. doi:10.1038/nclimate2079. Retrieved January 26, 2014.

- "Daily Updated Time series of Arctic sea ice area and extent derived from SSMI data provided by NERSC". Retrieved 14 September 2013.

Pollution Control Systems and Technologies

Environment-friendly technology encourages the use of long-lasting devices and cells for energy storage and consumption. It also focuses on re-use rather than throwing away otherwise non-bio-degradable material. Tools and techniques are an important component of any field of study. The following chapter elucidates the various tools and techniques that are related to pollution control.

Electrostatic Precipitator

Electrodes inside electrostatic precipitator

An electrostatic precipitator (ESP) is a filtration device that removes fine particles, like dust and smoke, from a flowing gas using the force of an induced electrostatic charge minimally impeding the flow of gases through the unit.

In contrast to wet scrubbers which apply energy directly to the flowing fluid medium, an ESP applies energy only to the particulate matter being collected and therefore is very efficient in its consumption of energy (in the form of electricity).

Collection electrode of electrostatic precipitator in waste incineration plant

Invention of the Electrostatic Precipitator

The first use of corona discharge to remove particles from an aerosol was by Hohlfeld in 1824. However, it was not commercialized until almost a century later.

In 1907 Frederick Gardner Cottrell, a professor of chemistry at the University of California, Berkeley, applied for a patent on a device for charging particles and then collecting them through electrostatic attraction—the first electrostatic precipitator. Cottrell first applied the device to the collection of sulphuric acid mist and lead oxide fumes emitted from various acid-making and smelting activities. Wine-producing vineyards in northern California were being adversely affected by the lead emissions.

At the time of Cottrell's invention, the theoretical basis for operation was not understood. The operational theory was developed later in Germany, with the work of Walter Deutsch and the formation of the Lurgi company.

Cottrell used proceeds from his invention to fund scientific research through the creation of a foundation called Research Corporation in 1912, to which he assigned the patents. The intent of the organization was to bring inventions made by educators (such as Cottrell) into the commercial world for the benefit of society at large. The operation of Research Corporation is funded by royalties paid by commercial firms after commercialization occurs. Research Corporation has provided vital funding to many scientific projects: Goddard's rocketry experiments, Lawrence's cyclotron, production methods for vitamins A and B_1, among many others.

By a decision of the US Supreme Court, the Corporation had to be split into several entities. The Research Corporation was separated from two commercial firms making the hardware: Research-Cottrell Inc. (operating east of the Mississippi River) and Western Precipitation (operating in the western states). The Research Corporation continues to be active to this day, and the two companies formed to commercialize the invention for industrial and utility applications are still in business as well.

Electrophoresis is the term used for migration of gas-suspended charged particles in a direct-current electrostatic field. Traditional CRT television sets tend to accumulate dust on the screen because of this phenomenon (a CRT is a direct-current machine operating at about 35 kilovolts).

Plate Precipitator

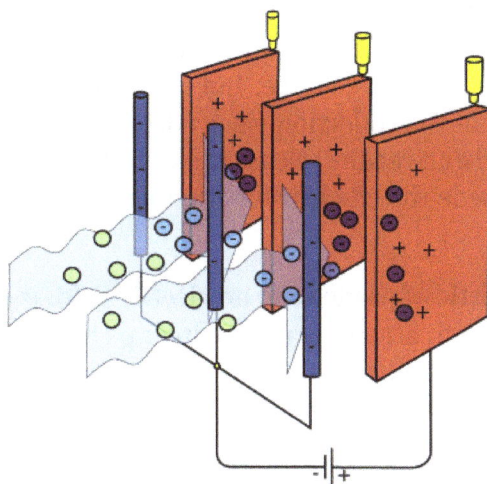

Conceptual diagram of an electrostatic precipitator

The most basic precipitator contains a row of thin vertical wires, and followed by a stack of large flat metal plates oriented vertically, with the plates typically spaced about 1 cm to 18 cm apart, depending on the application. The air stream flows horizontally through the spaces between the wires, and then passes through the stack of plates.

A negative voltage of several thousand volts is applied between wire and plate. If the applied voltage is high enough, an electric corona discharge ionizes the air around the electrodes, which then ionizes the particles in the air stream.

The ionized particles, due to the electrostatic force, are diverted towards the grounded plates. Particles build up on the collection plates and are removed from the air stream.

A two-stage design (separate charging section ahead of collecting section) has the benefit of minimizing ozone production, which would adversely affect health of personnel working in enclosed spaces. For shipboard engine rooms where gearboxes generate an oil mist, two-stage ESP's are used to clean the air, improving the operating environment and preventing buildup of flammable oil fog accumulations. Collected oil is returned to the gear lubricating system.

Collection Efficiency (R)

Precipitator performance is very sensitive to two particulate properties: 1) Electrical resistivity; and 2) Particle size distribution. These properties can be measured economically and accurately in the laboratory, using standard tests. Resistivity can be determined as a function of temperature in accordance with IEEE Standard 548. This test is conducted in an air environment containing a specified moisture concentration. The test is run as a function of ascending or descending temperature, or both. Data is acquired using an average ash layer electric field of 4 kV/cm. Since relatively low applied voltage is used and no sulfuric acid vapor is present in the test environment, the values obtained indicate the maximum ash resistivity.

In an ESP, where particle charging and discharging are key functions, resistivity is an important factor that significantly affects collection efficiency. While resistivity is an important phenomenon in the inter-electrode region where most particle charging takes place, it has a particularly important effect on the dust layer at the collection electrode where discharging occurs. Particles that exhibit high resistivity are difficult to charge. But once charged, they do not readily give up their acquired charge on arrival at the collection electrode. On the other hand, particles with low resistivity easily become charged and readily release their charge to the grounded collection plate. Both extremes in resistivity impede the efficient functioning of ESPs. ESPs work best under normal resistivity conditions.

Resistivity, which is a characteristic of particles in an electric field, is a measure of a particle's resistance to transferring charge (both accepting and giving up charges). Resistivity is a function of a particle's chemical composition as well as flue gas operating conditions such as temperature and moisture. Particles can have high, moderate (normal), or low resistivity.

Bulk resistivity is defined using a more general version of Ohm's Law, as given in Equation (1) below:

$$\vec{E} = \rho \vec{j} \tag{1}$$

Where:

E is the Electric field strength (V/cm);

j is the Current density (A/cm^2); and

ρ is the Resistivity (Ohm-cm)

A better way of displaying this would be to solve for resistivity as a function of applied voltage and current, as given in Equation (2) below:

$$\rho = \frac{AV}{Il} \tag{2}$$

Where:

ρ = Resistivity (Ohm-cm)

V = The applied DC potential, (Volts);

I = The measured current, (Amperes);

l = The ash layer thickness, (cm); and

A = The current measuring electrode face area, (cm^2).

Resistivity is the electrical resistance of a dust sample 1.0 cm^2 in cross-sectional area, 1.0 cm thick, and is recorded in units of ohm-cm. A method for measuring resistivity will be described in this article. The table below, gives value ranges for low, normal, and high resistivity.

Resistivity	Range of Measurement
Low	between 10^4 and 10^7 ohm-cm
Normal	between 10^7 and 2×10^{10} ohm-cm
High	above 2×10^{10} ohm-cm

Dust Layer Resistivity

Resistivity affects electrical conditions in the dust layer by a potential electric field (voltage drop) being formed across the layer as negatively charged particles arrive at its surface and leak their electrical charges to the collection plate. At the metal surface of the electrically grounded collection plate, the voltage is zero, whereas at the outer surface of the dust layer, where new particles and ions are arriving, the electrostatic voltage caused by the gas ions can be quite high. The strength of this electric field depends on the resistivity and thickness of the dust layer.

In high-resistivity dust layers, the dust is not sufficiently conductive, so electrical charges have difficulty moving through the dust layer. Consequently, electrical charges accumulate on and beneath the dust layer surface, creating a strong electric field.

Voltages can be greater than 10,000 volts. Dust particles with high resistivities are held too strongly to the plate, making them difficult to remove and causing rapping problems.

In low resistivity dust layers, the corona current is readily passed to the grounded collection electrode. Therefore, a relatively weak electric field, of several thousand volts, is maintained across the dust layer. Collected dust particles with low resistivity do not adhere strongly enough to the collection plate. They are easily dislodged and become re-entrained in the gas stream.

The electrical conductivity of a bulk layer of particles depends on both surface and volume factors. Volume conduction, or the motions of electrical charges through the interiors of particles, depends mainly on the composition and temperature of the particles. In the higher temperature regions, above 500 °F (260 °C), volume conduction controls the conduction mechanism. Volume conduction also involves ancillary factors, such as compression of the particle layer, particle size and shape, and surface properties.

Volume conduction is represented in the figures as a straight-line at temperatures above 500 °F (260 °C). At temperatures below about 450 °F (230 °C), electrical charges begin to flow across surface moisture and chemical films adsorbed onto the particles. Surface conduction begins to lower the resistivity values and bend the curve downward at temperatures below 500 °F (260 °C).

These films usually differ both physically and chemically from the interiors of the particles owing to adsorption phenomena. Theoretical calculations indicate that moisture films only a few molecules thick are adequate to provide the desired surface conductivity. Surface conduction on particles is closely related to surface-leakage currents occurring on electrical insulators, which have been extensively studied. An interesting practical application of surface-leakage is the determination of dew point by measurement of the current between adjacent electrodes mounted on a glass surface. A sharp rise in current signals the formation of a moisture film on the glass. This method has been used effectively for determining the marked rise in dew point, which occurs when small amounts of sulfuric acid vapor are added to an atmosphere (commercial Dewpoint Meters are available on the market).

The following discussion of normal, high, and low resistivity applies to ESPs operated in a dry state; resistivity is not a problem in the operation of wet ESPs because of the moisture concentration in the ESP. The relationship between moisture content and resistivity is explained later in this work.

Normal Resistivity

As stated above, ESPs work best under normal resistivity conditions. Particles with normal resistivity do not rapidly lose their charge on arrival at the collection electrode. These particles slowly leak their charge to grounded plates and are retained on the collection plates by intermolecular adhesive and cohesive forces. This allows a particulate layer to be built up and then dislodged from the plates by rapping. Within the range of normal dust resistivity (between 10^7 and 2×10^{10} ohm-cm), fly ash is collected more easily than dust having either low or high resistivity.

High Resistivity

If the voltage drop across the dust layer becomes too high, several adverse effects can occur. First, the high voltage drop reduces the voltage difference between the discharge electrode and collection electrode, and thereby reduces the electrostatic field strength used to drive the gas ion-charged particles over to the collected dust layer. As the dust layer builds up, and the electrical charges accumulate on the surface of the dust layer, the voltage difference between the discharge and collection electrodes decreases. The migration velocities of small particles are especially affected by the reduced electric field strength.

Another problem that occurs with high resistivity dust layers is called back corona. This occurs when the potential drop across the dust layer is so great that corona discharges begin to appear in the gas that is trapped within the dust layer. The dust layer breaks down electrically, producing small holes or craters from which back corona discharges occur. Positive gas ions are generated within the dust layer and are accelerated toward the "negatively charged" discharge electrode. The positive ions reduce some of the negative charges on the dust layer and neutralize some of the negative ions on the "charged particles" heading toward the collection electrode. Disruptions of the normal corona process greatly reduce the ESP's collection efficiency, which in severe cases, may fall below 50% . When back corona is present, the dust particles build up on the electrodes forming a layer of insulation. Often this can not be repaired without bringing the unit offline.

The third, and generally most common problem with high resistivity dust is increased electrical sparking. When the sparking rate exceeds the "set spark rate limit," the automatic controllers limit the operating voltage of the field. This causes reduced particle charging and reduced migration velocities toward the collection electrode. High resistivity can generally be reduced by doing the following:

- Adjusting the temperature;

- Increasing moisture content;

- Adding conditioning agents to the gas stream;

- Increasing the collection surface area; and

- Using hot-side precipitators (occasionally and with foreknowledge of sodium depletion).

Thin dust layers and high-resistivity dust especially favor the formation of back corona craters. Severe back corona has been observed with dust layers as thin as 0.1 mm, but a dust layer just over one particle thick can reduce the sparking voltage by 50%. The most marked effects of back corona on the current-voltage characteristics are:

1. Reduction of the spark over voltage by as much as 50% or more;

2. Current jumps or discontinuities caused by the formation of stable back-corona craters; and

3. Large increase in maximum corona current, which just below spark over corona gap may be several times the normal current.

The Figure below and to the left shows the variation in resistivity with changing gas temperature for six different industrial dusts along with three coal-fired fly ashes. The Figure on the right illustrates resistivity values measured for various chemical compounds that were prepared in the laboratory.

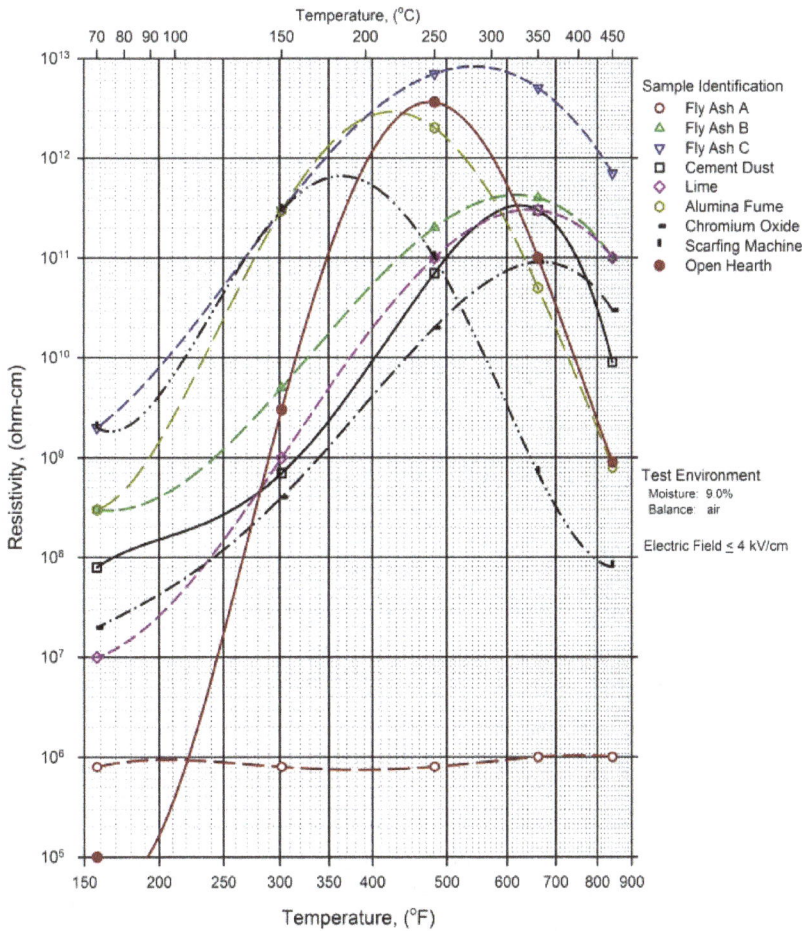

Resistivity Values of Representative Dusts and Fumes From Industrial Plants

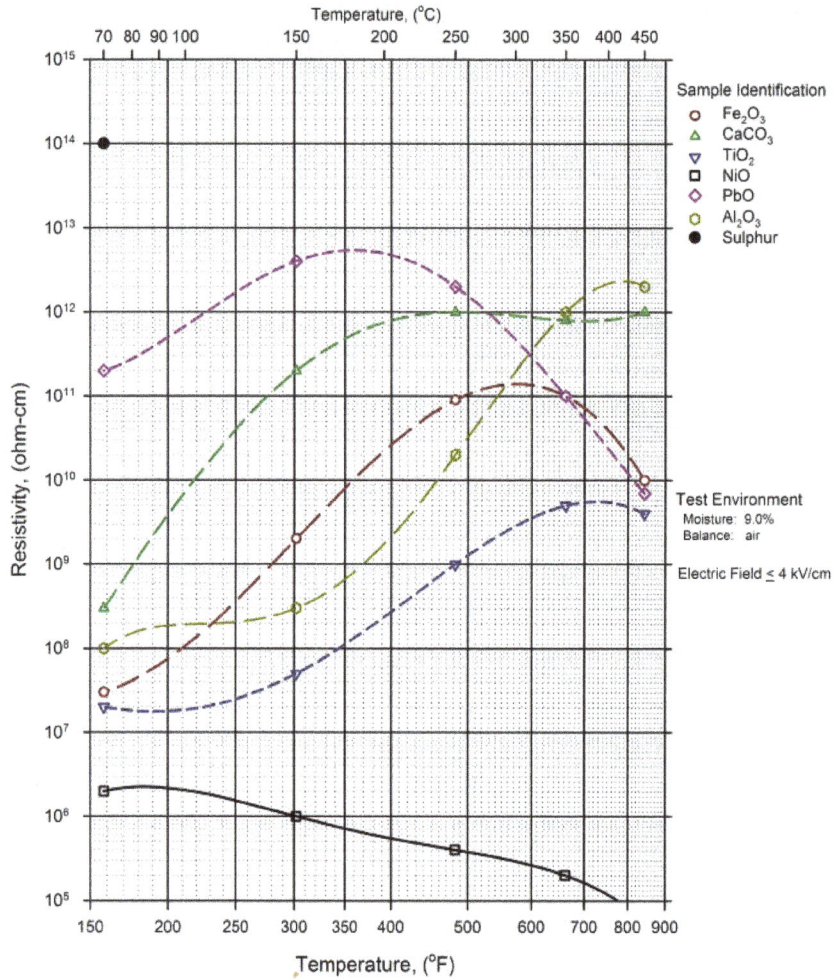

Resistivity Values of Various Chemicals and Reagents as a Function of Temperature

Results for Fly Ash A (in the figure to the left) were acquired in the ascending temperature mode. These data are typical for a moderate to high combustibles content ash. Data for Fly Ash B are from the same sample, acquired during the descending temperature mode.

The differences between the ascending and descending temperature modes are due to the presence of unburned combustibles in the sample. Between the two test modes, the samples are equilibrated in dry air for 14 hours (overnight) at 850 °F (450 °C). This overnight annealing process typically removes between 60% and 90% of any unburned combustibles present in the samples. Exactly how carbon works as a charge carrier is not fully understood, but it is known to significantly reduce the resistivity of a dust.

Carbon can act, at first, like a high resistivity dust in the precipitator. Higher voltages can be required in order for corona generation to begin. These higher voltages can be problematic for the TR-Set controls. The problem lies in onset of corona causing large amounts of current to surge through the (low resistivity) dust layer. The controls sense this surge as a spark. As precipitators are operated in spark-limiting mode, power is terminated and the corona generation cycle re-initiates. Thus, lower power (current) readings are noted with relatively high voltage readings.

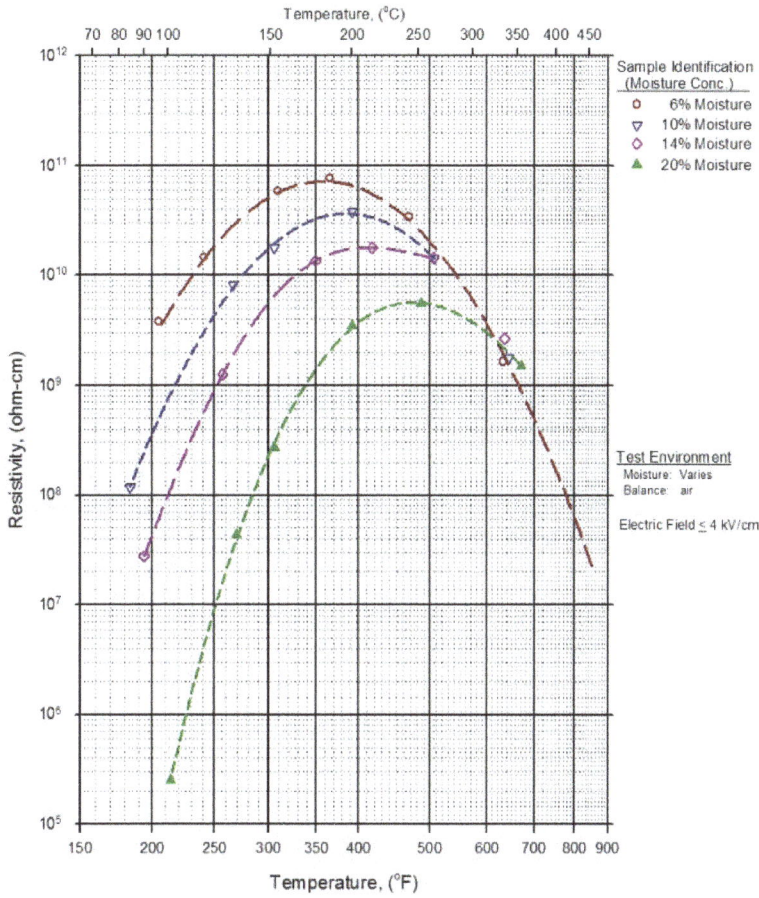

Resistivity Measured as a Function of Temperature in Varying Moisture Concentrations (Humidity)

The same thing is believed to occur in laboratory measurements. Parallel plate geometry is used in laboratory measurements without corona generation. A stainless steel cup holds the sample. Another stainless steel electrode weight sits on top of the sample (direct contact with the dust layer). As voltage is increased from small amounts (e.g. 20 V), no current is measured. Then, a threshold voltage level is reached. At this level, current surges through the sample... so much so that the voltage supply unit can trip off. After removal of the unburned combustibles during the above-mentioned annealing procedure, the descending temperature mode curve shows the typical inverted "V" shape one might expect.

Low Resistivity

Particles that have low resistivity are difficult to collect because they are easily charged (very conductive) and rapidly lose their charge on arrival at the collection electrode. The particles take on the charge of the collection electrode, bounce off the plates, and become re-entrained in the gas stream. Thus, attractive and repulsive electrical forces that are normally at work at normal and higher resistivities are lacking, and the binding forces to the plate are considerably lessened. Examples of low-resistivity dusts are unburned carbon in fly ash and carbon black.

If these conductive particles are coarse, they can be removed upstream of the precipitator by using a device such as a cyclone mechanical collector.

The addition of liquid ammonia (NH_3) into the gas stream as a conditioning agent has found wide use in recent years. It is theorized that ammonia reacts with H_2SO_4 contained in the flue gas to form an ammonium sulfate compound that increases the cohesivity of the dust. This additional cohesivity makes up for the loss of electrical attraction forces.

The table below summarizes the characteristics associated with low, normal and high resistivity dusts.

The moisture content of the flue gas stream also affects particle resistivity. Increasing the moisture content of the gas stream by spraying water or injecting steam into the duct work preceding the ESP lowers the resistivity. In both temperature adjustment and moisture conditioning, one must maintain gas conditions above the dew point to prevent corrosion problems in the ESP or downstream equipment. The figure to the right shows the effect of temperature and moisture on the resistivity of a cement dust. As the percentage of moisture in the gas stream increases from 6 to 20%, the resistivity of the dust dramatically decreases. Also, raising or lowering the temperature can decrease cement dust resistivity for all the moisture percentages represented.

The presence of SO_3 in the gas stream has been shown to favor the electrostatic precipitation process when problems with high resistivity occur. Most of the sulfur content in the coal burned for combustion sources converts to SO_2. However, approximately 1% of the sulfur converts to SO_3. The amount of SO_3 in the flue gas normally increases with increasing sulfur content of the coal. The resistivity of the particles decreases as the sulfur content of the coal increases.

Resistivity	Range of Measurement	Precipitator Characteristics
Low	between 10^4 and 10^7 ohm-cm	1. Normal operating voltage and current levels unless dust layer is thick enough to reduce plate clearances and cause higher current levels. 2. Reduced electrical force component retaining collected dust, vulnerable to high reentrainment losses. 3. Negligible voltage drop across dust layer. 4. Reduced collection performance due to (2)
Normal	between 10^7 and 2×10^{10} ohm-cm	1. Normal operating voltage and current levels. 2. Negligible voltage drop across dust layer. 3. Sufficient electrical force component retaining collected dust. 4. High collection performance due to (1), (2) and (3)
Marginal to High	between 2×10^{10} and 10^{12} ohm-cm	1. Reduced operating voltage and current levels with high spark rates. 2. Significant voltage loss across dust layer. 3. Moderate electrical force component retaining collected dust. 4. Reduced collection performance due to (1) and (2)

High	above 10^{12} ohm-cm	1. Reduced operating voltage levels; high operating current levels if power supply controller is not operating properly.
		2. Very significant voltage loss across dust layer.
		3. High electrical force component retaining collected dust.
		4. Seriously reduced collection performance due to (1), (2) and probably back corona.

Other conditioning agents, such as sulfuric acid, ammonia, sodium chloride, and soda ash (sometimes as raw trona), have also been used to reduce particle resistivity. Therefore, the chemical composition of the flue gas stream is important with regard to the resistivity of the particles to be collected in the ESP. The table below lists various conditioning agents and their mechanisms of operation.

Conditioning Agent	Mechanism(s) of Action
Sulfur Trioxide and/or Sulfuric Acid	1. Condensation and adsorption on fly ash surfaces. 2. may also increase cohesiveness of fly ash. 3. Reduces resistivity
Ammonia	1. Mechanism is not clear, various ones proposed; 2. Modifies resistivity. 3. Increases ash cohesiveness. 4. Enhances space charge effect.
Ammonium Sulfate	1. Little is known about the mechanism; claims are made for the following: 2. Modifies resistivity (depends upon injection temperature). 3. Increases ash cohesiveness. 4. Enhances space charge effect. 5. Experimental data lacking to substantiate which of these is predominant.
Triethylamine	1. Particle agglomeration claimed; no supporting data.
Sodium Compounds	1. Natural conditioner if added with coal. 2. Resistivity modifier if injected into gas stream.
Compounds of Transition Metals	1. Postulated that they catalyze oxidation of SO_2 to SO_3; no definitive tests with fly ash to verify this postulation.
Potassium Sulfate and Sodium Chloride	1. In cement and lime kiln ESPs: 2. Resistivity modifiers in the gas stream. 3. NaCl - natural conditioner when mixed with coal.

If injection of ammonium sulfate occurs at a temperature greater than about 600 °F (320 °C), dissociation into ammonia and sulfur trioxide results. Depending on the ash, SO_2 may preferentially interact with fly ash as SO_3 conditioning. The remainder recombines with ammonia to add to the space charge as well as increase cohesiveness of the ash.

More recently, it has been recognized that a major reason for loss of efficiency of the electrostatic precipitator is due to particle buildup on the charging wires in addition to the collection plates (Davidson and McKinney, 1998). This is easily remedied by making sure that the wires themselves are cleaned at the same time that the collecting plates are cleaned.

Sulfuric acid vapor (SO_3) enhances the effects of water vapor on surface conduction. It is physically adsorbed within the layer of moisture on the particle surfaces. The effects of relatively small amounts of acid vapor can be seen in the figure below and to the right.

The inherent resistivity of the sample at 300 °F (150 °C) is 5×10^{12} ohm-cm. An equilibrium concentration of just 1.9 ppm sulfuric acid vapor lowers that value to about 7×10^9 ohm-cm.

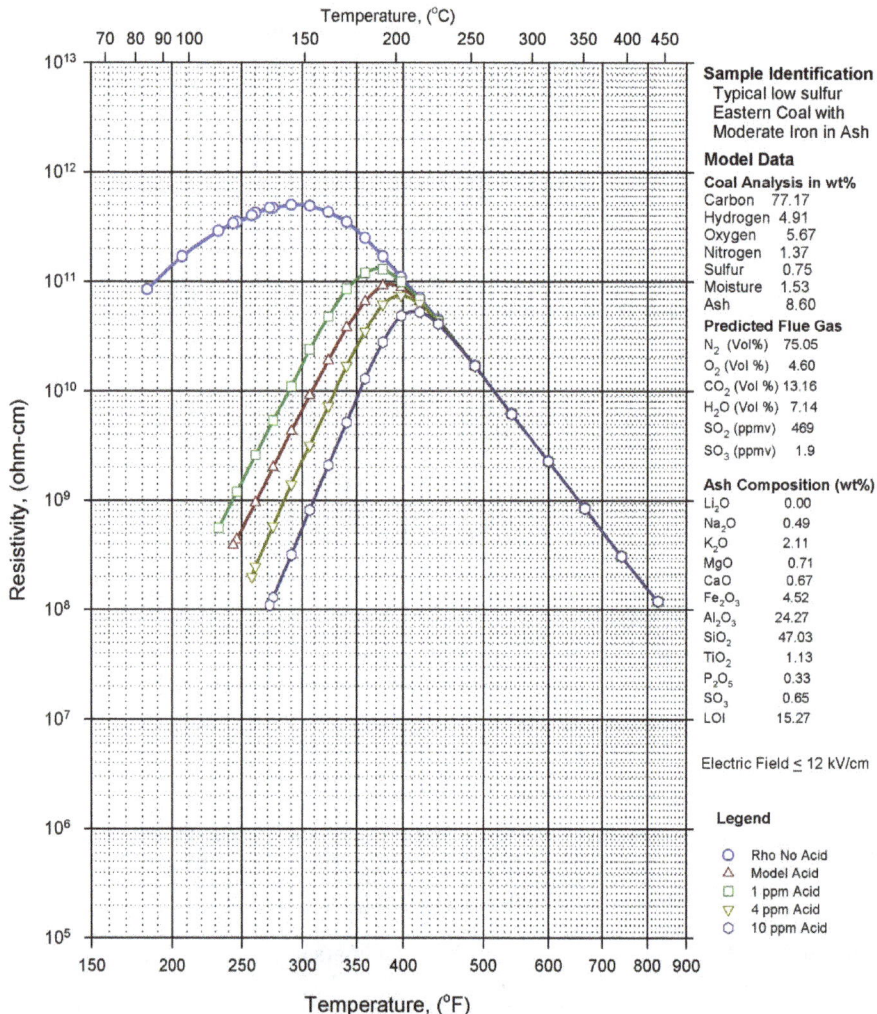

Resistivity Modeled As A Function of Environmental Conditions - Especially Sulfuric Acid Vapor

Modern Industrial Electrostatic Precipitators

A smokestack at coal-fired Hazelwood Power Station in Victoria,
Australia emits brown smoke when its ESP is shut down

ESPs continue to be excellent devices for control of many industrial particulate emissions, including smoke from electricity-generating utilities (coal and oil fired), salt cake collection from black liquor boilers in pulp mills, and catalyst collection from fluidized bed catalytic cracker units in oil refineries to name a few. These devices treat gas volumes from several hundred thousand ACFM to 2.5 million ACFM (1,180 m³/s) in the largest coal-fired boiler applications. For a coal-fired boiler the collection is usually performed downstream of the air preheater at about 160 °C (320 °F) which provides optimal resistivity of the coal-ash particles. For some difficult applications with low-sulfur fuel hot-end units have been built operating above 370 °C (698 °F).

The original parallel plate–weighted wire design has evolved as more efficient (and robust) discharge electrode designs were developed, today focusing on rigid (pipe-frame) discharge electrodes to which many sharpened spikes are attached (barbed wire), maximizing corona production. Transformer-rectifier systems apply voltages of 50–100 kV at relatively high current densities. Modern controls, such as an automatic voltage control, minimize electric sparking and prevent arcing (sparks are quenched within 1/2 cycle of the TR set), avoiding damage to the components. Automatic plate-rapping systems and hopper-evacuation systems remove the collected particulate matter while on line, theoretically allowing ESPs to stay in continuous operation for years at a time.

Electrostatic Air Sampling

Electrostatic precipitators can be used to sample airborne particles or aerosol for analysis purposes. Precipitator designs with a liquid counterelectrode can be used to sample biological particles, e.g. viruses, directly into a small liquid volume to reduce unnecessary sample dilution.

Wet Electrostatic Precipitator

A wet electrostatic precipitator (WESP or wet ESP) operates with water vapor saturated air streams (100% relative humidity). WESPs are commonly used to remove liquid droplets such as sulfuric acid mist from industrial process gas streams. The WESP is also commonly used where the gases are high in moisture content, contain combustible particulate, or have particles that are sticky in nature.

The preferred and most modern type of WESP is a downflow tubular design. This design allows the collected moisture and particulate to form a moving slurry that helps to keep the collection surfaces clean. Plate style and upflow design WESPs are very unreliable and should not be used in applications where particulate is sticky in nature.

Consumer-oriented Electrostatic Air Cleaners

A portable electrostatic air cleaner marketed to consumers

Portable electrostatic air cleaner with cover removed, showing collector plates

Plate precipitators are commonly marketed to the public as air purifier devices or as a permanent replacement for furnace filters, but all have the undesirable attribute of being somewhat messy to clean. A negative side-effect of electrostatic precipitation devices is the potential production of toxic ozone and NO_x. However, electrostatic precipitators offer benefits over other air purifications technologies, such as HEPA filtration, which require expensive filters and can become "production sinks" for many harmful forms of bacteria.

With electrostatic precipitators, if the collection plates are allowed to accumulate large amounts of particulate matter, the particles can sometimes bond so tightly to the metal plates that vigorous washing and scrubbing may be required to completely clean the collection plates. The close spacing of the plates can make thorough cleaning difficult, and the stack of plates often cannot be easily disassembled for cleaning. One solution, suggested by several manufacturers, is to wash the collector plates in a dishwasher.

Some consumer precipitation filters are sold with special soak-off cleaners, where the entire plate array is removed from the precipitator and soaked in a large container overnight, to help loosen the tightly bonded particulates.

A study by the Canada Mortgage and Housing Corporation testing a variety of forced-air furnace filters found that ESP filters provided the best, and most cost-effective means of cleaning air using a forced-air system.

The first portable electrostatic air filter systems for homes was marketed in 1954 by Raytheon.

Scrubber

Scrubber systems are a diverse group of air pollution control devices that can be used to remove some particulates and/or gases from industrial exhaust streams. The first air scrubber was designed to remove carbon dioxide from the air of an early submarine, the Ictineo I, a role for which they continue to be used to this day. Traditionally, the term "scrubber" has referred to pollution control devices that use liquid to wash unwanted pollutants from a gas stream. Recently, the term is also used to describe systems that inject a dry reagent or slurry into a dirty exhaust stream to "wash out" acid gases. Scrubbers are one of the primary devices that control gaseous emissions, especially acid gases. Scrubbers can also be used for heat recovery from hot gases by flue-gas condensation.

There are several methods to remove toxic or corrosive compounds from exhaust gas and neutralize it.

Combustion

Combustion is sometimes the cause for harmful exhaust, but, in many cases, combustion may also be used for exhaust gas cleaning if the temperature is high enough and enough oxygen is available.

Wet Scrubbing

The exhaust gases of combustion may contain substances considered harmful to the environment, and the scrubber may remove or neutralize those. A wet scrubber is used to clean air, fuel gas or other gases of various pollutants and dust particles. Wet scrubbing works via the contact of target compounds or particulate matter with the scrubbing solution. Solutions may simply be water (for dust) or solutions of reagents that specifically target certain compounds.

Process exhaust gas can also contain water-soluble toxic and/or corrosive gases like hydrochloric acid (HCl) or ammonia (NH_3). These can be removed very well by a wet scrubber.

Removal efficiency of pollutants is improved by increasing residence time in the scrubber or by the increase of surface area of the scrubber solution by the use of a spray nozzle, packed towers or an aspirator. Wet scrubbers may increase the proportion of water in the gas, resulting in a visible stack plume, if the gas is sent to a stack.

Wet scrubbers can also be used for heat recovery from hot gases by flue-gas condensation. In this mode, termed a condensing scrubber, water from the scrubber drain is circulated through a cooler to the nozzles at the top of the scrubber. The hot gas enters the scrubber at the bottom. If the gas temperature is above the water dew point, it is initially cooled by evaporation of water drops. Further cooling cause water vapors to condense, adding to the amount of circulating water.

The condensation of water release significant amounts of low temperature heat (more than 2 gigajoules (560 kWh) per ton of water), that can be recovered by the cooler for e.g. district heating purposes.

Excess condensed water must continuously be removed from the circulating water.

The gas leaves the scrubber at its dew point, so even though significant amounts of water may have been removed from the cooled gas, it is likely to leave a visible stack plume of water vapor.

Dry Scrubbing

A dry or semi-dry scrubbing system, unlike the wet scrubber, does not saturate the flue gas stream that is being treated with moisture. In some cases no moisture is added, while in others only the amount of moisture that can be evaporated in the flue gas without condensing is added. Therefore, dry scrubbers generally do not have a stack steam plume or wastewater handling/disposal requirements. Dry scrubbing systems are used to remove acid gases (such as SO_2 and HCl) primarily from combustion sources.

There are a number of dry type scrubbing system designs. However, all consist of two main sections or devices: a device to introduce the acid gas sorbent material into the gas stream and a particulate matter control device to remove reaction products, excess sorbent material as well as any particulate matter already in the flue gas.

Dry scrubbing systems can be categorized as dry sorbent injectors (DSIs) or as spray dryer absorbers (SDAs). Spray dryer absorbers are also called semi-dry scrubbers or spray dryers.

Dry scrubbing systems are often used for the removal of odorous and corrosive gases from wastewater treatment plant operations. The medium used is typically an activated alumina compound impregnated with materials to handle specific gases such as hydrogen sulfide. Media used can be mixed together to offer a wide range of removal for other odorous compounds such as methyl mercaptans, aldehydes, volatile organic compounds, dimethyl sulfide, and dimethyl disulfide.

Dry sorbent injection involves the addition of an alkaline material (usually hydrated lime, soda ash, or sodium bicarbonate) into the gas stream to react with the acid gases. The sorbent can be injected directly into several different locations: the combustion process, the flue gas duct (ahead of the particulate control device), or an open reaction chamber (if one exists). The acid gases react with the alkaline sorbents to form solid salts which are removed in the particulate control device. These simple systems can achieve only limited acid gas (SO_2 and HCl) removal efficiencies. Higher collection efficiencies can be achieved by increasing the flue gas humidity (i.e., cooling using water spray). These devices have been used on medical waste incinerators and a few municipal waste combustors.

In spray dryer absorbers, the flue gases are introduced into an absorbing tower (dryer) where the gases are contacted with a finely atomized alkaline slurry. Acid gases are absorbed by the slurry mixture and react to form solid salts which are removed by the particulate control device. The heat of the flue gas is used to evaporate all the water droplets, leaving a non-saturated flue gas to exit the absorber tower. Spray dryers are capable of achieving high (80+%) acid gas removal efficiencies. These devices have been used on industrial and utility boilers and municipal waste incinerators.

Absorber

Many chemicals can be removed from exhaust gas also by using absorber material. The flue gas is passed through a cartridge which is filled with one or several absorber materials and has been adapted to the chemical properties of the components to be removed. This type of scrubber is sometimes also called dry scrubber. The absorber material has to be replaced after its surface is saturated.

Mercury Removal

Mercury is a highly toxic element commonly found in coal and municipal waste. Wet scrubbers are only effective for removal of soluble mercury species, such as oxidized mercury, Hg^{2+}. Mercury vapor in its elemental form, Hg^0, is insoluble in the scrubber slurry and not removed. Therefore, additional process of Hg^0 conversion is required to complete mercury capture. Usually addition of the halogens to the flue gas are used for this purpose. The type of coal burned as well as the presence of a selective catalytic reduction unit both affect the ratio of elemental to oxidized mercury in the flue gas and thus the degree to which the mercury is removed.

In July 2015, one study found that some mercury scrubbers installed on coal power plants inadvertently capture PAH (polycyclic aromatic hydrocarbons) emissions as well.

Scrubber Waste Products

One side effect of scrubbing is that the process only moves the unwanted substance from the exhaust gases into a liquid solution, solid paste or powder form. This must be disposed of safely, if it can not be reused.

For example, mercury removal results in a waste product that either needs further processing to extract the raw mercury, or must be buried in a special hazardous wastes landfill that prevents the mercury from seeping out into the environment.

As an example of reuse, limestone-based scrubbers in coal-fired power plants can produce a synthetic gypsum of sufficient quality that can be used to manufacture drywall and other industrial products.

Bacteria Spread

Poorly maintained scrubbers have the potential to spread disease-causing bacteria. The problem is a result of inadequate cleaning. For example, the cause of a 2005 outbreak of Legionnaires' disease in Norway was just a few infected scrubbers. The outbreak caused 10 deaths and more than 50 cases of infection.

Wet Scrubber

The term wet scrubber describes a variety of devices that remove pollutants from a furnace flue gas or from other gas streams. In a wet scrubber, the polluted gas stream is brought into contact with the scrubbing liquid, by spraying it with the liquid, by forcing it through a pool of liquid, or by some other contact method, so as to remove the pollutants.

Design

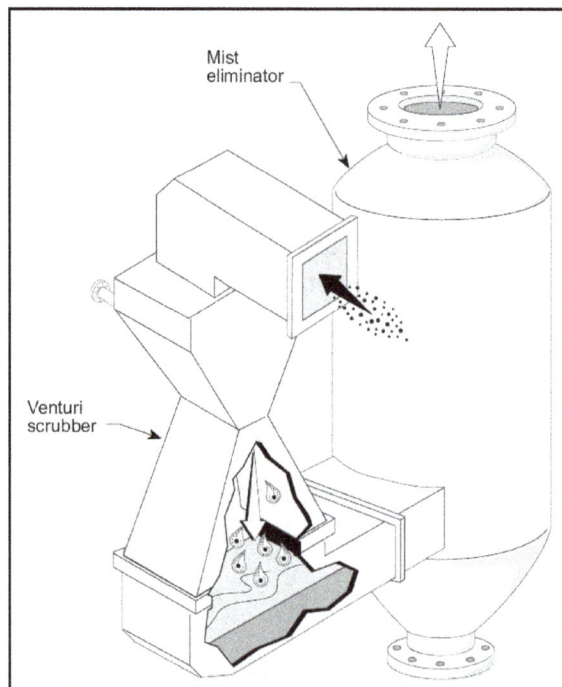

A venturi scrubber design. The mist eliminator for a venturi scrubber is often a separate device called a cyclonic separator

The design of wet scrubbers or any air pollution control device depends on the industrial process conditions and the nature of the air pollutants involved. Inlet gas characteristics and dust properties (if particles are present) are of primary importance. Scrubbers can be designed to collect particulate matter and/or gaseous pollutants. The versatility of wet scrubbers allow them to be built in numerous configurations, all designed to provide good contact between the liquid and polluted gas stream.

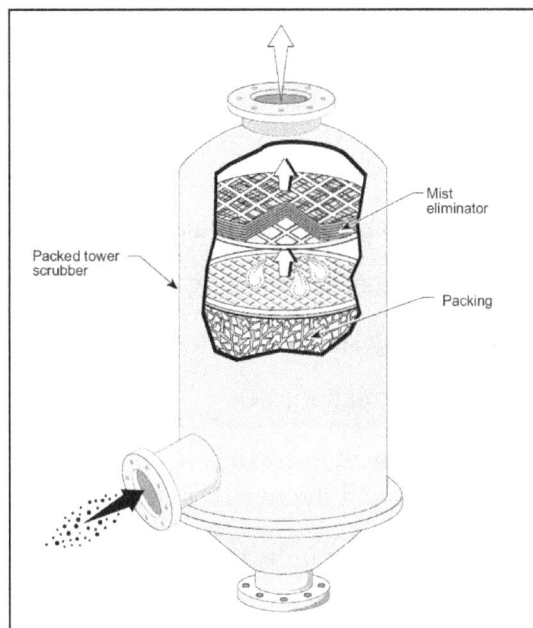

A packed bed tower design where the mist eliminator is built into the top of the structure. Various tower designs exist

Wet scrubbers remove dust particles by *capturing* them in liquid droplets. The droplets are then collected, the liquid *dissolving* or *absorbing* the pollutant gases. Any droplets that are in the scrubber inlet gas must be separated from the outlet gas stream by means of another device referred to as a mist eliminator or entrainment separator (these terms are interchangeable). Also, the resultant scrubbing liquid must be treated prior to any ultimate discharge or being reused in the plant.

A wet scrubber's ability to collect small particles is often directly proportional to the power input into the scrubber. Low energy devices such as spray towers are used to collect particles larger than 5 micrometers. To obtain high efficiency removal of 1 micrometer (or less) particles generally requires high energy devices such as venturi scrubbers or augmented devices such as condensation scrubbers. Additionally, a properly designed and operated entrainment separator or mist eliminator is important to achieve high removal efficiencies. The greater the number of liquid droplets that are not captured by the mist eliminator, the higher the potential emission levels.

Wet scrubbers that remove gaseous pollutants are referred to as *absorbers*. Good gas-to-liquid contact is essential to obtain high removal efficiencies in absorbers. A number of wet scrubber designs are used to remove gaseous pollutants, with the packed tower and the plate tower being the most common.

If the gas stream contains both particle matter and gases, wet scrubbers are generally the only single air pollution control device that can remove both pollutants. Wet scrubbers can achieve high removal efficiencies for either particles or gases and, in some instances, can achieve a high removal efficiency for both pollutants in the same system. However, in many cases, the best operating conditions for particles collection are the poorest for gas removal.

In general, obtaining high simultaneous gas and particulate removal efficiencies requires that one of them be easily collected (i.e., that the gases are very soluble in the liquid or that the particles are large and readily captured) or by the use of a scrubbing reagent such as lime or sodium hydroxide.

Advantages and Disadvantages

For particulate control, wet scrubbers (also referred to as wet collectors) are evaluated against fabric filters and electrostatic precipitators (ESPs). Some *advantages* of wet scrubbers over these devices are as follows:

- Wet scrubbers have the ability to handle high temperatures and moisture.

- In wet scrubbers, flue gases are cooled, resulting in smaller overall size of equipment.

- Wet scrubbers can remove both gases and particulate matter.

- Wet scrubbers can neutralize corrosive gases.

Some *disadvantages* of wet scrubbers include corrosion, the need for entrainment separation or mist removal to obtain high efficiencies and the need for treatment or reuse of spent liquid.

Wet scrubbers have been used in a variety of industries such as acid plants, fertilizer plants, steel mills, asphalt plants, and large power plants.

Relative advantages and disadvantages of wet scrubbers compared to other control devices	
Advantages	**Disadvantages**
• Small space requirements: Scrubbers reduce the temperature and volume of the unsaturated exhaust stream. Therefore, vessel sizes, including fans and ducts downstream, are smaller than those of other control devices. Smaller sizes result in lower capital costs and more flexibility in site location of the scrubber. • No secondary dust sources: Once particulate matter is collected, it cannot escape from hoppers or during transport. • Handles high-temperature, high-humidity gas streams: No temperature limits or condensation problems can occur as in baghouses or ESPs. • Minimal fire and explosion hazards: Various dry dusts are flammable. Using water eliminates the possibility of explosions. • Ability to collect both gases and particulate matter.	• Corrosion problems: Water and dissolved pollutants can form highly corrosive acid solutions. Proper construction materials are very important. Also, wet-dry interface areas can result in corrosion. • High power requirements: High collection efficiencies for particulate matter are attainable only at high pressure drops, resulting in high operating costs. • Water-disposal problems: Settling ponds or sludge clarifiers may be needed to meet waste-water regulations. • Difficult product recovery: Dewatering and drying of scrubber sludge make recovery of any dust for reuse very expensive and difficult.

Components

Wet scrubber systems generally consist of the following components:

- Ductwork and fan system

- A saturation chamber (optional)

- Scrubbing vessel

- Entrainment separator or mist eliminator

- Pumping (and possible recycle system)

- Spent scrubbing liquid treatment and/or reuse system

- An exhaust stack

A typical wet scrubbing process can be described as follows:

- Hot flue gas from a furnace enters a saturator (if present) where gases are cooled and humidified prior to entering the scrubbing area. The saturator removes a small percentage of the particulate matter present in the flue gas.

- Next, the gas enters a venturi scrubber where approximately half of the gases are removed. Venturi scrubbers have a minimum particle removal efficiency of 95%.

- The gas flows through a second scrubber, a packed bed absorber, where the rest of the gases (and particulate matter) are collected.

- An entrainment separator or mist eliminator removes any liquid droplets that may have become entrained in the flue gas.

- A recirculation pump moves some of the spent scrubbing liquid back to the venturi scrubber where it is recycled and the remainder is sent to a treatment system.

- Treated scrubbing liquid is recycled back to the saturator and the packed bed absorber.

- Fans and ductwork move the flue gas stream through the system and eventually out the stack.

Categorization

Since wet scrubbers vary greatly in complexity and method of operation, devising categories into which all of them neatly fit is extremely difficult. Scrubbers for particle collection are usually categorized by the gas-side pressure drop of the system. Gas-side pressure drop refers to the pressure difference, or pressure drop, that occurs as the exhaust gas is pushed or pulled through the scrubber, disregarding the pressure that would be used for pumping or spraying the liquid into the scrubber.

Scrubbers may be classified *by pressure drop* as follows:

- *Low-energy scrubbers* have pressure drops of less than 12.7 cm (5 in) of water.

- *Medium-energy scrubbers* have pressure drops between 12.7 and 38.1 cm (5 and 15 in) of water.

- *High-energy scrubbers* have pressure drops greater than 37.1 cm (15 in) of water.

However, most scrubbers operate over a wide range of pressure drops, depending on their specific application, thereby making this type of categorization difficult.

Another way to classify wet scrubbers is by their *use* - to primarily collect either particulates or gaseous pollutants. Again, this distinction is not always clear since scrubbers can often be used to remove both types of pollutants.

Wet scrubbers can also be categorized by the manner in which the gas and liquid phases are brought into contact. Scrubbers are designed to use power, or energy, from the gas stream or the liquid stream, or some other method to bring the pollutant gas stream into contact with the liquid. These categories are given in Table 2.

Categories of wet collectors by energy source used for contact	
Wet collector	**Energy source used for gas-liquid contact**
• Gas-phase contacting	• Gas stream
• Liquid-phase contacting	• Liquid stream
• Wet film	• Liquid and gas streams
• Combination	• Energy source:
o Liquid phase and gas phase	o Liquid and gas streams
o Mechanically aided	o Mechanically driven rotor

Material of Construction and Design

Corrosion can be a prime problem associated with chemical industry scrubbing systems. Fibre-reinforced plastic and dual laminates are often used as most dependable materials of construction.

Sewage Treatment

Wastewater treatment plant in Massachusetts, United States

Sewage treatment is the process of removing contaminants from wastewater, primarily from household sewage. It includes physical, chemical, and biological processes to remove these contaminants and produce environmentally safe treated wastewater (or treated effluent). A by-product of sewage treatment is usually a semi-solid waste or slurry, called sewage sludge, that has to undergo further treatment before being suitable for disposal or land application.

Sewage treatment may also be referred to as wastewater treatment, although the latter is a broader term which can also be applied to purely industrial wastewater. For most cities, the sewer system will also carry a proportion of industrial effluent to the sewage treatment plant which has usually received pretreatment at the factories themselves to reduce the pollutant load. If the sewer system is a combined sewer then it will also carry urban runoff (stormwater) to the sewage treatment plant.

Terminology

The term "sewage treatment plant" (or "sewage treatment works" in some countries) is nowadays often replaced with the term "wastewater treatment plant".

Sewage can be treated close to where the sewage is created, which may be called a "decentralized" system or even an "on-site" system (in septic tanks, biofilters or aerobic treatment systems). Alternatively, sewage can be collected and transported by a network of pipes and pump stations to a municipal treatment plant. This is called a "centralized" system, although the borders between decentralized and centralized can be variable. For this reason, the terms "semi-decentralized" and "semi-centralized" are also being used.

Origins of Sewage

Sewage is generated by residential, institutional, commercial and industrial establishments. It includes household waste liquid from toilets, baths, showers, kitchens, and sinks draining into sewers. In many areas, sewage also includes liquid waste from industry and commerce. The separation and draining of household waste into greywater and blackwater is becoming more common in the developed world, with treated greywater being permitted to be used for watering plants or recycled for flushing toilets.

Sewage Mixing with Rainwater

Sewage may include stormwater runoff or urban runoff. Sewerage systems capable of handling storm water are known as combined sewer systems. This design was common when urban sewerage systems were first developed, in the late 19th and early 20th centuries. Combined sewers require much larger and more expensive treatment facilities than sanitary sewers. Heavy volumes of storm runoff may overwhelm the sewage treatment system, causing a spill or overflow. Sanitary sewers are typically much smaller than combined sewers, and they are not designed to transport stormwater. Backups of raw sewage can occur if excessive infiltration/inflow (dilution by stormwater and/or groundwater) is allowed into a sanitary sewer system. Communities that have urbanized in the mid-20th century or later generally have built separate systems for sewage (sanitary sewers) and stormwater, because precipitation causes widely varying flows, reducing sewage treatment plant efficiency.

As rainfall travels over roofs and the ground, it may pick up various contaminants including soil particles and other sediment, heavy metals, organic compounds, animal waste, and oil and grease. Some jurisdictions require stormwater to receive some level of treatment before being discharged directly into waterways. Examples of treatment processes used for stormwater include retention basins, wetlands, buried vaults with various kinds of media filters, and vortex separators (to remove coarse solids).

Industrial Effluent

In highly regulated developed countries, industrial effluent usually receives at least pretreatment if not full treatment at the factories themselves to reduce the pollutant load, before discharge to the sewer. This process is called industrial wastewater treatment. The same does not apply to many developing countries where industrial effluent is more likely to enter the sewer if it exists, or even the receiving water body, without pretreatment.

Industrial wastewater may contain pollutants which cannot be removed by conventional sewage treatment. Also, variable flow of industrial waste associated with production cycles may upset the population dynamics of biological treatment units, such as the activated sludge process.

Process Steps

Overview

Sewage collection and treatment is typically subject to local, state and federal regulations and standards.

Treating wastewater has the aim to produce an effluent that will do as little harm as possible when discharged to the surrounding environment, thereby preventing pollution compared to releasing untreated wastewater into the environment.

Sewage treatment generally involves three stages, called primary, secondary and tertiary treatment.

- *Primary treatment* consists of temporarily holding the sewage in a quiescent basin where heavy solids can settle to the bottom while oil, grease and lighter solids float to the surface. The settled and floating materials are removed and the remaining liquid may be discharged or subjected to secondary treatment. Some sewage treatment plants that are connected to a combined sewer system have a bypass arrangement after the primary treatment unit. This means that during very heavy rainfall events, the secondary and tertiary treatment systems can be bypassed to protect them from hydraulic overloading, and the mixture of sewage and stormwater only receives primary treatment.

- *Secondary treatment* removes dissolved and suspended biological matter. Secondary treatment is typically performed by indigenous, water-borne micro-organisms in a managed habitat. Secondary treatment may require a separation process to remove the micro-organisms from the treated water prior to discharge or tertiary treatment.

- *Tertiary treatment* is sometimes defined as anything more than primary and secondary treatment in order to allow rejection into a highly sensitive or fragile ecosystem (estuaries, low-flow rivers, coral reefs,...). Treated water is sometimes disinfected chemically or physically (for example, by lagoons and microfiltration) prior to discharge into a stream, river, bay, lagoon or wetland, or it can be used for the irrigation of a golf course, green way or park. If it is sufficiently clean, it can also be used for groundwater recharge or agricultural purposes.

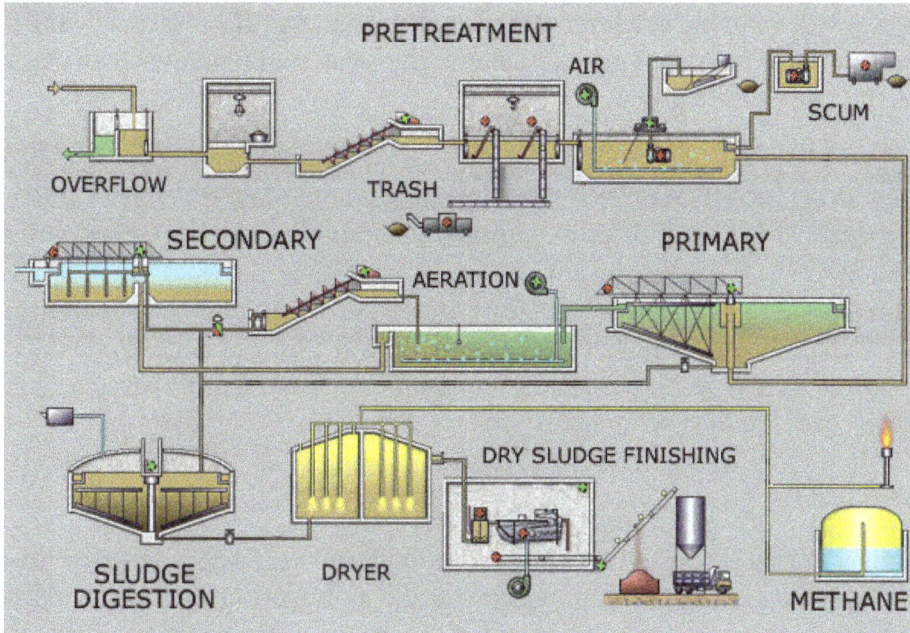

Simplified process flow diagram for a typical large-scale treatment plant

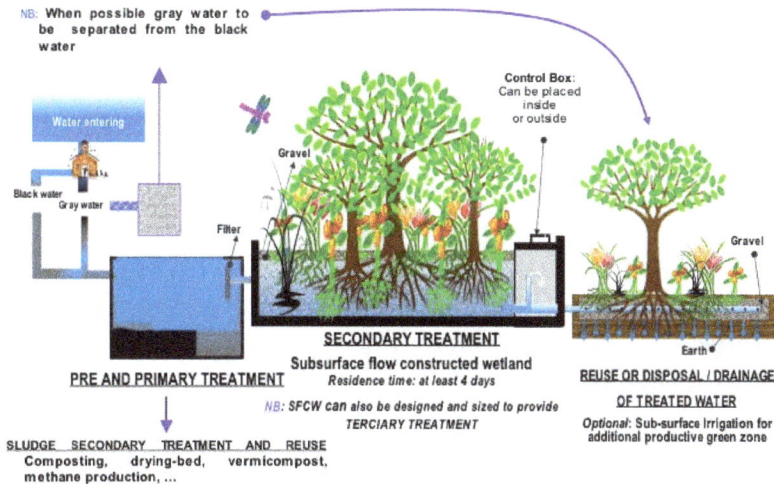

Process flow diagram for a typical treatment plant via subsurface flow constructed wetlands (SFCW)

Pretreatment

Pretreatment removes all materials that can be easily collected from the raw sewage before they damage or clog the pumps and sewage lines of primary treatment clarifiers. Objects commonly removed during pretreatment include trash, tree limbs, leaves, branches, and other large objects.

The influent in sewage water passes through a bar screen to remove all large objects like cans, rags, sticks, plastic packets etc. carried in the sewage stream. This is most commonly done with an automated mechanically raked bar screen in modern plants serving large populations, while in smaller or less modern plants, a manually cleaned screen may be used. The raking action of a mechanical bar screen is typically paced according to the accumulation on the bar screens and/or flow rate. The solids are collected and later disposed in a landfill, or incinerated. Bar screens or mesh screens of varying sizes may be used to optimize solids removal. If gross solids are not removed, they become entrained in pipes and moving parts of the treatment plant, and can cause substantial damage and inefficiency in the process.

Grit Removal

Pretreatment may include a sand or grit channel or chamber, where the velocity of the incoming sewage is adjusted to allow the settlement of sand, grit, stones, and broken glass. These particles are removed because they may damage pumps and other equipment. For small sanitary sewer systems, the grit chambers may not be necessary, but grit removal is desirable at larger plants. Grit chambers come in 3 types: horizontal grit chambers, aerated grit chambers and vortex grit chambers. The process is called sedimentation.

Flow Equalization

Clarifiers and mechanized secondary treatment are more efficient under uniform flow conditions. Equalization basins may be used for temporary storage of diurnal or wet-weather flow peaks. Basins provide a place to temporarily hold incoming sewage during plant maintenance and a means of diluting and distributing batch discharges of toxic or high-strength waste which might otherwise inhibit biological secondary treatment (including portable toilet waste, vehicle holding tanks, and septic tank pumpers). Flow equalization basins require variable discharge control, typically include provisions for bypass and cleaning, and may also include aerators. Cleaning may be easier if the basin is downstream of screening and grit removal.

Fat and Grease Removal

In some larger plants, fat and grease are removed by passing the sewage through a small tank where skimmers collect the fat floating on the surface. Air blowers in the base of the tank may also be used to help recover the fat as a froth. Many plants, however, use primary clarifiers with mechanical surface skimmers for fat and grease removal.

Primary Treatment

In the primary sedimentation stage, sewage flows through large tanks, commonly called "pre-settling basins", "primary sedimentation tanks" or "primary clarifiers". The tanks are used to settle sludge while grease and oils rise to the surface and are skimmed off. Primary settling tanks are usually equipped with mechanically driven scrapers that continually drive the collected sludge towards a hopper in the base of the tank where it is pumped to sludge treatment facilities. Grease and oil from the floating material can sometimes be recovered for saponification (soap making).

Primary treatment tanks in Oregon, USA.

Secondary Treatment

Secondary treatment is designed to substantially degrade the biological content of the sewage which are derived from human waste, food waste, soaps and detergent. The majority of municipal plants treat the settled sewage liquor using aerobic biological processes. To be effective, the biota require both oxygen and food to live. The bacteria and protozoa consume biodegradable soluble organic contaminants (e.g. sugars, fats, organic short-chain carbon molecules, etc.) and bind much of the less soluble fractions into floc. Secondary treatment systems are classified as *fixed-film* or *suspended-growth* systems.

- Fixed-film or attached growth systems include trickling filters, bio-towers, and rotating biological contactors, where the biomass grows on media and the sewage passes over its surface. The fixed-film principle has further developed into Moving Bed Biofilm Reactors (MBBR) and Integrated Fixed-Film Activated Sludge (IFAS) processes. An MBBR system typically requires a smaller footprint than suspended-growth systems.

- Suspended-growth systems include activated sludge, where the biomass is mixed with the sewage and can be operated in a smaller space than trickling filters that treat the same amount of water. However, fixed-film systems are more able to cope with drastic changes in the amount of biological material and can provide higher removal rates for organic material and suspended solids than suspended growth systems.

Secondary Sedimentation

Some secondary treatment methods include a secondary clarifier to settle out and separate biological floc or filter material grown in the secondary treatment bioreactor.

Secondary clarifier at a rural treatment plant.

List of Process Types

- Activated sludge

- Aerated lagoon

- Aerobic granulation

- Constructed wetland

- Membrane bioreactor

- Rotating biological contactor

- Sequencing batch reactor

- Trickling filter

To use less space, treat difficult waste, and intermittent flows, a number of designs of hybrid treatment plants have been produced. Such plants often combine at least two stages of the three main treatment stages into one combined stage. In the UK, where a large number of wastewater treatment plants serve small populations, package plants are a viable alternative to building a large structure for each process stage. In the US, package plants are typically used in rural areas, highway rest stops and trailer parks.

Tertiary Treatment

The purpose of tertiary treatment is to provide a final treatment stage to further improve the effluent quality before it is discharged to the receiving environment (sea, river, lake, wet lands, ground, etc.). More than one tertiary treatment process may be used at any treatment plant. If disinfection is practised, it is always the final process. It is also called "effluent polishing."

Filtration

Sand filtration removes much of the residual suspended matter. Filtration over activated carbon, also called *carbon adsorption,* removes residual toxins.

Lagoons or Ponds

A sewage treatment plant and lagoon in Everett, Washington, United States.

Lagoons or ponds provide settlement and further biological improvement through storage in large man-made ponds or lagoons. These lagoons are highly aerobic and colonization by native macrophytes, especially reeds, is often encouraged. Small filter feeding invertebrates such as *Daphnia* and species of *Rotifera* greatly assist in treatment by removing fine particulates.

Biological Nutrient Removal

Biological nutrient removal (BNR) is regarded by some as a type of secondary treatment process, and by others as a tertiary (or "advanced") treatment process.

Wastewater may contain high levels of the nutrients nitrogen and phosphorus. Excessive release to the environment can lead to a buildup of nutrients, called eutrophication, which can in turn encourage the overgrowth of weeds, algae, and cyanobacteria (blue-green algae). This may cause an algal bloom, a rapid growth in the population of algae. The algae numbers are unsustainable and eventually most of them die. The decomposition of the algae by bacteria uses up so much of the oxygen in the water that most or all of the animals die, which creates more organic matter for the bacteria to decompose. In addition to causing deoxygenation, some algal species produce toxins that contaminate drinking water supplies. Different treatment processes are required to remove nitrogen and phosphorus.

Nitrogen Removal

Nitrogen is removed through the biological oxidation of nitrogen from ammonia to nitrate (nitrification), followed by denitrification, the reduction of nitrate to nitrogen gas. Nitrogen gas is released to the atmosphere and thus removed from the water.

Nitrification itself is a two-step aerobic process, each step facilitated by a different type of bacteria. The oxidation of ammonia (NH_3) to nitrite (NO_2^-) is most often facilitated by *Nitrosomonas* spp. ("nitroso" referring to the formation of a nitroso functional group). Nitrite oxidation to nitrate (NO_3^-), though traditionally believed to be facilitated by *Nitrobacter* spp. (nitro referring the formation of a nitro functional group), is now known to be facilitated in the environment almost exclusively by *Nitrospira* spp.

Denitrification requires anoxic conditions to encourage the appropriate biological communities to form. It is facilitated by a wide diversity of bacteria. Sand filters, lagooning and reed beds can all be used to reduce nitrogen, but the activated sludge process (if designed well) can do the job the most easily. Since denitrification is the reduction of nitrate to dinitrogen (molecular nitrogen) gas, an electron donor is needed. This can be, depending on the waste water, organic matter (from feces), sulfide, or an added donor like methanol. The sludge in the anoxic tanks (denitrification tanks) must be mixed well (mixture of recirculated mixed liquor, return activated sludge [RAS], and raw influent) e.g. by using submersible mixers in order to achieve the desired denitrification.

Sometimes the conversion of toxic ammonia to nitrate alone is referred to as tertiary treatment.

Over time, different treatment configurations have evolved as denitrification has become more sophisticated. An initial scheme, the Ludzack-Ettinger Process, placed an anoxic treatment zone before the aeration tank and clarifier, using the return activated sludge (RAS) from the clarifier as a nitrate source. Influent wastewater (either raw or as effluent from primary clarification) serves as the electron source for the facultative bacteria to metabolize carbon, using the inorganic nitrate as a source of oxygen instead of dissolved molecular oxygen. This denitrification scheme was naturally limited to the amount of soluble nitrate present in the RAS. Nitrate reduction was limited because RAS rate is limited by the performance of the clarifier.

The "Modified Ludzak-Ettinger Process" (MLE) is an improvement on the original concept, for it recycles mixed liquor from the discharge end of the aeration tank to the head of the anoxic tank to provide a consistent source of soluble nitrate for the facultative bacteria. In this instance, raw wastewater continues to provide the electron source, and sub-surface mixing maintains the bacteria in contact with both electron source and soluble nitrate in the absence of dissolved oxygen.

Many sewage treatment plants use centrifugal pumps to transfer the nitrified mixed liquor from the aeration zone to the anoxic zone for denitrification. These pumps are often referred to as *Internal Mixed Liquor Recycle* (IMLR) pumps. IMLR may be 200% to 400% the flow rate of influent wastewater (Q.) This is in addition to Return Activated Sludge (RAS) from secondary clarifiers, which may be 100% of Q. (Therefore, the hydraulic capacity of the tanks in such a system should handle at least 400% of annual average design flow (AADF.) At times, the raw or primary effluent wastewater must be carbon-supplemented by the addition of methanol, acetate, or simple food waste (molasses, whey, plant starch) to improve the treatment efficiency. These carbon additions should be accounted for in the design of a treatment facility's organic loading.

Further modifications to the MLE were to come: Bardenpho and Biodenipho processes include additional anoxic and oxidative processes to further polish the conversion of nitrate ion to molecular nitrogen gas. Use of an anaerobic tank following the initial anoxic process allows for luxury uptake of phosphorus by bacteria, thereby biologically reducing orthophosphate ion in the treated wastewater. Even newer improvements, such as Anammox Process, interrupt the formation of nitrate at the nitrite stage of nitrification, shunting nitrite-rich mixed liquor activated sludge to treatment where nitrite is then converted to molecular nitrogen gas, saving energy, alkalinity, and secondary carbon sourcing. Anammox™ (ANaerobic AMMonia OXidation) works by artificially extending detention time and preserving denitrifiying bacteria through the use of substrate added to the mixed liquor and continuously recycled from it prior to secondary clarification. Many other proprietary schemes are being deployed, including DEMON™, Sharon-ANAMMOX™, ANITA-Mox™, and DeAmmon™. The bacteria Brocadia anammoxidans can remove ammonium from waste water through anaerobic oxidation of ammonium to hydrazine, a form of rocket fuel.

Phosphorus Removal

Every adult human excretes between 200 and 1000 grams of phosphorus annually. Studies of United States sewage in the late 1960s estimated mean per capita contributions of 500 grams in urine and feces, 1000 grams in synthetic detergents, and lesser variable amounts used as corrosion and scale control chemicals in water supplies. Source control via alternative detergent formulations has subsequently reduced the largest contribution, but the content of urine and feces will remain unchanged. Phosphorus removal is important as it is a limiting nutrient for algae growth in many fresh water systems. It is also particularly important for water reuse systems where high phosphorus concentrations may lead to fouling of downstream equipment such as reverse osmosis.

Phosphorus can be removed biologically in a process called enhanced biological phosphorus removal. In this process, specific bacteria, called polyphosphate-accumulating organisms (PAOs), are selectively enriched and accumulate large quantities of phosphorus within their cells (up to 20 percent of their mass). When the biomass enriched in these bacteria is separated from the treated water, these biosolids have a high fertilizer value.

Phosphorus removal can also be achieved by chemical precipitation, usually with salts of iron (e.g. ferric chloride), aluminum (e.g. alum), or lime. This may lead to excessive sludge production as hydroxides precipitates and the added chemicals can be expensive. Chemical phosphorus removal requires significantly smaller equipment footprint than biological removal, is easier to operate and is often more reliable than biological phosphorus removal. Another method for phosphorus removal is to use granular laterite.

Once removed, phosphorus, in the form of a phosphate-rich sewage sludge, may be dumped in a landfill or used as fertilizer. In the latter case, the treated sewage sludge is also sometimes referred to as biosolids.

Disinfection

The purpose of disinfection in the treatment of waste water is to substantially reduce the number of microorganisms in the water to be discharged back into the environment for the later use of drinking, bathing, irrigation, etc. The effectiveness of disinfection depends on the quality of the water being treated (e.g., cloudiness, pH, etc.), the type of disinfection being used, the disinfectant dosage (concentration and time), and other environmental variables. Cloudy water will be treated less successfully, since solid matter can shield organisms, especially from ultraviolet light or if contact times are low. Generally, short contact times, low doses and high flows all militate against effective disinfection. Common methods of disinfection include ozone, chlorine, ultraviolet light, or sodium hypochlorite. Chloramine, which is used for drinking water, is not used in the treatment of waste water because of its persistence. After multiple steps of disinfection, the treated water is ready to be released back into the water cycle by means of the nearest body of water or agriculture. Afterwards, the water can be transferred to reserves for everyday human uses.

Chlorination remains the most common form of waste water disinfection in North America due to its low cost and long-term history of effectiveness. One disadvantage is that chlorination of residual organic material can generate chlorinated-organic compounds that may be carcinogenic or harmful to the environment. Residual chlorine or chloramines may also be capable of chlorinating organic material in the natural aquatic environment. Further, because residual chlorine is toxic to aquatic species, the treated effluent must also be chemically dechlorinated, adding to the complexity and cost of treatment.

Ultraviolet (UV) light can be used instead of chlorine, iodine, or other chemicals. Because no chemicals are used, the treated water has no adverse effect on organisms that later consume it, as may be the case with other methods. UV radiation causes damage to the genetic structure of bacteria, viruses, and other pathogens, making them incapable of reproduction. The key disadvantages of UV disinfection are the need for frequent lamp maintenance and replacement and the need for a highly treated effluent to ensure that the target microorganisms are not shielded from the UV radiation (i.e., any solids present in the treated effluent may protect microorganisms from the UV light). In the United Kingdom, UV light is becoming the most common means of disinfection because of the concerns about the impacts of chlorine in chlorinating residual organics in the wastewater and in chlorinating organics in the receiving water. Some sewage treatment systems in Canada and the US also use UV light for their effluent water disinfection.

Ozone (O_3) is generated by passing oxygen (O_2) through a high voltage potential resulting in a third oxygen atom becoming attached and forming O_3. Ozone is very unstable and reactive and oxidizes most organic material it comes in contact with, thereby destroying many pathogenic microorganisms. Ozone is considered to be safer than chlorine because, unlike chlorine which has to be stored on site (highly poisonous in the event of an accidental release), ozone is generated on-site as needed. Ozonation also produces fewer disinfection by-products than chlorination. A disadvantage of ozone disinfection is the high cost of the ozone generation equipment and the requirements for special operators.

Fourth Treatment Stage

Micropollutants such as pharmaceuticals, ingredients of household chemicals, chemicals used in small businesses or industries, environmental persistent pharmaceutical pollutant (EPPP) or pesticides may not be eliminated in the conventional treatment process (primary, secondary and tertiary treatment) and therefore lead to water pollution. Although concentrations of those substances and their decompostion products are quite low, there is still a chance to harm aquatic organisms. For pharmaceuticals, the following substances have been identified as "toxicologically relevant": substances with endocrine disrupting effects, genotoxic substances and substances that enhance the development of bacterial resistances. They mainly belong to the group of environmental persistent pharmaceutical pollutants. Techniques for elimination of micropollutants via a fourth treatment stage during sewage treatment are being tested in Germany, Switzerland and the Netherlands. However, since those techniques are still costly, they are not yet applied on a regular basis. Such process steps mainly consist of activated carbon filters that adsorb the micropollutants. Ozone can also be applied as an oxidative method. Also the use of enzymes such as the enzyme laccase is under investigation. A new concept which could provide an energy-efficient treatment of micropollutants could be the use of laccase secreting fungi cultivated at a wastewater treatment plant to degrade micropollutants and at the same time to provide enzymes at a cathode of a microbial biofuel cells. Microbial biofuel cells are investigated for their property to treat organic matter in wastewater.

To reduce pharmaceuticals in water bodies, also "source control" measures are under investigation, such as innovations in drug development or more responsible handling of drugs.

Odor Control

Odors emitted by sewage treatment are typically an indication of an anaerobic or "septic" condition. Early stages of processing will tend to produce foul-smelling gases, with hydrogen sulfide being most common in generating complaints. Large process plants in urban areas will often treat the odors with carbon reactors, a contact media with bio-slimes, small doses of chlorine, or circulating fluids to biologically capture and metabolize the noxious gases. Other methods of odor control exist, including addition of iron salts, hydrogen peroxide, calcium nitrate, etc. to manage hydrogen sulfide levels.

High-density solids pumps are suitable for reducing odors by conveying sludge through hermetic closed pipework.

Energy Requirements

For conventional sewage treatment plants, around 30 percent of the annual operating costs is usually required for energy. The energy requirements vary with type of treatment process as well as wastewater load. For example, constructed wetlands have a lower energy requirement than activated sludge plants, as less energy is required for the aeration step. Sewage treatment plants that produce biogas in their sewage sludge treatment process with anaerobic digestion can produce enough energy to meet most of the energy needs of the sewage treatment plant itself.

In conventional secondary treatment processes, most of the electricity is used for aeration, pumping systems and equipment for the dewatering and drying of sewage sludge. Advanced wastewater treatment plants, e.g. for nutrient removal, require more energy than plants that only achieve primary or secondary treatment.

Sludge Treatment and Disposal

The sludges accumulated in a wastewater treatment process must be treated and disposed of in a safe and effective manner. The purpose of digestion is to reduce the amount of organic matter and the number of disease-causing microorganisms present in the solids. The most common treatment options include anaerobic digestion, aerobic digestion, and composting. Incineration is also used, albeit to a much lesser degree.

Sludge treatment depends on the amount of solids generated and other site-specific conditions. Composting is most often applied to small-scale plants with aerobic digestion for mid-sized operations, and anaerobic digestion for the larger-scale operations.

The sludge is sometimes passed through a so-called pre-thickener which de-waters the sludge. Types of pre-thickeners include centrifugal sludge thickeners rotary drum sludge thickeners and belt filter presses. Dewatered sludge may be incinerated or transported offsite for disposal in a landfill or use as an agricultural soil amendment.

Environment Aspects

The outlet of the Karlsruhe sewage treatment plant flows into the Alb.

Many processes in a wastewater treatment plant are designed to mimic the natural treatment processes that occur in the environment, whether that environment is a natural water body or the ground. If not overloaded, bacteria in the environment will consume organic contaminants, although this will reduce the levels of oxygen in the water and may significantly change the overall ecology of the receiving water. Native bacterial populations feed on the organic contaminants, and the numbers of disease-causing microorganisms are reduced by natural environmental conditions such as predation or exposure to ultraviolet radiation. Consequently, in cases where the receiving environment provides a high level of dilution, a high degree of wastewater treatment may not be required. However, recent evidence has demonstrated that very low levels of specific contaminants in wastewater, including hormones (from animal husbandry and residue from human hormonal contraception methods) and synthetic materials such as phthalates that mimic hormones in their action, can have an unpredictable adverse impact on the natural biota and potentially on humans if the water is re-used for drinking water. In the US and EU, uncontrolled discharges of wastewater to the environment are not permitted under law, and strict water quality requirements are to be met, as clean drinking water is essential. A significant threat in the coming decades will be the increasing uncontrolled discharges of wastewater within rapidly developing countries.

Effects on Biology

Sewage treatment plants can have multiple effects on nutrient levels in the water that the treated sewage flows into. These nutrients can have large effects on the biological life in the water in contact with the effluent. Stabilization ponds (or sewage treatment ponds) can include any of the following:

- Oxidation ponds, which are aerobic bodies of water usually 1–2 meters in depth that receive effluent from sedimentation tanks or other forms of primary treatment.

  - Dominated by algae

- Polishing ponds are similar to oxidation ponds but receive effluent from an oxidation pond or from a plant with an extended mechanical treatment.

  - Dominated by zooplankton

- Facultative lagoons, raw sewage lagoons, or sewage lagoons are ponds where sewage is added with no primary treatment other than coarse screening. These ponds provide effective treatment when the surface remains aerobic; although anaerobic conditions may develop near the layer of settled sludge on the bottom of the pond.

- Anaerobic lagoons are heavily loaded ponds.

  - Dominated by bacteria

- Sludge lagoons are aerobic ponds, usually 2 to 5 meters in depth, that receive anaerobically digested primary sludge, or activated secondary sludge under water.

  - Upper layers are dominated by algae

Phosphorus limitation is a possible result from sewage treatment and results in flagellate-dominated plankton, particularly in summer and fall.

A phytoplankton study found high nutrient concentrations linked to sewage effluents. High nutrient concentration leads to high chlorophyll a concentrations, which is a proxy for primary production in marine environments. High primary production means high phytoplankton populations and most likely high zooplankton populations, because zooplankton feed on phytoplankton. However, effluent released into marine systems also leads to greater population instability.

The planktonic trends of high populations close to input of treated sewage is contrasted by the bacterial trend. In a study of *Aeromonas* spp. in increasing distance from a wastewater source, greater change in seasonal cycles was found the furthest from the effluent. This trend is so strong that the furthest location studied actually had an inversion of the *Aeromonas* spp. cycle in comparison to that of fecal coliforms. Since there is a main pattern in the cycles that occurred simultaneously at all stations it indicates seasonal factors (temperature, solar radiation, phytoplankton) control of the bacterial population. The effluent dominant species changes from *Aeromonas caviae* in winter to *Aeromonas sobria* in the spring and fall while the inflow dominant species is *Aeromonas caviae*, which is constant throughout the seasons.

Treated Sewage Reuse

With suitable technology, it is possible to reuse sewage effluent for drinking water, although this is usually only done in places with limited water supplies, such as Windhoek and Singapore.

In Israel, about 50 percent of agricultural water use (total use was 1 billion cubic metres in 2008) is provided through reclaimed sewer water. Future plans call for increased use of treated sewer water as well as more desalination plants.

Sewage Treatment in Developing Countries

Few reliable figures exist on the share of the wastewater collected in sewers that is being treated in the world. A global estimate by UNDP and UN-Habitat is that 90% of all wastewater generated is released into the environment untreated. In many developing countries the bulk of domestic and industrial wastewater is discharged without any treatment or after primary treatment only.

In Latin America about 15 percent of collected wastewater passes through treatment plants (with varying levels of actual treatment). In Venezuela, a below average country in South America with respect to wastewater treatment, 97 percent of the country's sewage is discharged raw into the environment. In Iran, a relatively developed Middle Eastern country, the majority of Tehran's population has totally untreated sewage injected to the city's groundwater. However, the construction of major parts of the sewage system, collection and treatment, in Tehran is almost complete, and under development, due to be fully completed by the end of 2012. In Isfahan, Iran's third largest city, sewage treatment was started more than 100 years ago.

Only few cities in sub-Saharan Africa have sewer-based sanitation systems, let alone wastewater treatment plants, an exception being South Africa and – until the late 1990s- Zimbabwe. Instead, most urban residents in sub-Saharan Africa rely on on-site sanitation systems without sewers, such as septic tanks and pit latrines, and faecal sludge management in these cities is an enormous challenge.

History

Basic sewer systems were used for waste removal in ancient Mesopotamia, where vertical shafts carried the waste away into cesspools. Similar systems existed in the Indus Valley civilization in modern-day India and in Ancient Crete and Greece. In the Middle Ages the sewer systems built by the Romans fell into disuse and waste was collected into cesspools that were periodically emptied by workers known as 'rakers' who would often sell it as fertilizer to farmers outside the city.

Modern sewage systems were first built in the mid-nineteenth century as a reaction to the exacerbation of sanitary conditions brought on by heavy industrialization and urbanization. Due to the contaminated water supply, cholera outbreaks occurred in 1832, 1849 and 1855 in London, killing tens of thousands of people. This, combined with the Great Stink of 1858, when the smell of untreated human waste in the River Thames became overpowering, and the report into sanitation reform of the Royal Commissioner Edwin Chadwick, led to the Metropolitan Commission of Sewers appointing Sir Joseph Bazalgette to construct a vast underground sewage system for the safe removal of waste. Contrary to Chadwick's recommendations, Bazalgette's system, and others later built in Continental Europe, did not pump the sewage onto farm land for use as fertilizer; it was simply piped to a natural waterway away from population centres, and pumped back into the environment.

The Great Stink of 1858 stimulated research into the problem of sewage treatment. In this caricature in *The Times*, Michael Faraday reports to *Father Thames* on the state of the river.

Early Attempts

One of the first attempts at diverting sewage for use as a fertilizer in the farm was made by the cotton mill owner James Smith in the 1840s. He experimented with a piped distribution system initially proposed by James Vetch that collected sewage from his factory and pumped it into the outlying farms, and his success was enthusiastically followed by Edwin Chadwick and supported by organic chemist Justus von Liebig.

The idea was officially adopted by the Health of Towns Commission, and various schemes (known as sewage farms) were trialled by different municipalities over the next 50 years. At first, the heavier solids were channeled into ditches on the side of the farm and were covered over when full, but soon flat-bottomed tanks were employed as reservoirs for the sewage; the earliest patent was taken out by William Higgs in 1846 for "tanks or reservoirs in which the contents of sewers and drains from cities, towns and villages are to be collected and the solid animal or vegetable matters therein contained, solidified and dried..." Improvements to the design of the tanks included the introduction of the horizontal-flow tank in the 1850s and the radial-flow tank in 1905. These tanks had to be manually de-sludged periodically, until the introduction of automatic mechanical de-sludgers in the early 1900s.

The precursor to the modern septic tank was the cesspool in which the water was sealed off to prevent contamination and the solid waste was slowly liquified due to anaerobic action; it was invented by L.H Mouras in France in the 1860s. Donald Cameron, as City Surveyor for Exeter patented an improved version in 1895, which he called a 'septic tank'; septic having the meaning of 'bacterial'. These are still in worldwide use, especially in rural areas unconnected to large-scale sewage systems.

Chemical Treatment

Sir Edward Frankland, a distinguished chemist, who demonstrated the possibility of chemically treating sewage in the 1870s.

It was not until the late 19th century that it became possible to treat the sewage by chemically breaking it down through the use of microorganisms and removing the pollutants. Land treatment was also steadily becoming less feasible, as cities grew and the volume of sewage produced could no longer be absorbed by the farmland on the outskirts.

Sir Edward Frankland conducted experiments at the Sewage Farm in Croydon, England, during the 1870s and was able to demonstrate that filtration of sewage through porous gravel produced a nitrified effluent (the ammonia was converted into nitrate) and that the filter remained unclogged over long periods of time. This established the then revolutionary possibility of biological treatment of sewage using a contact bed to oxidize the waste. This concept was taken up by the chief chemist for the London Metropolitan Board of Works, William Libdin, in 1887:

> ...in all probability the true way of purifying sewage...will be first to separate the sludge, and then turn into neutral effluent... retain it for a sufficient period, during which time it should be fully aerated, and finally discharge it into the stream in a purified condition. This is indeed what is aimed at and imperfectly accomplished on a sewage farm.

From 1885 to 1891 filters working on this principle were constructed throughout the UK and the idea was also taken up in the US at the Lawrence Experiment Station in Massachusetts, where Frankland's work was confirmed. In 1890 the LES developed a 'trickling filter' that gave a much more reliable performance.

Contact beds were developed in Salford, Lancashire and by scientists working for the London City Council in the early 1890s. According to Christopher Hamlin, this was part of a conceptual revolution that replaced the philosophy that saw "sewage purification as the prevention of decomposition with one that tried to facilitate the biological process that destroy sewage naturally."

Contact beds were tanks containing the inert substance, such as stones or slate, that maximized the surface area available for the microbial growth to break down the sewage. The sewage was held in the tank until it was fully decomposed and it was then filtered out into the ground. This method quickly became widespread, especially in the UK, where it was used in Leicester, Sheffield, Manchester and Leeds. The bacterial bed was simultaneously developed by Joseph Corbett as Borough Engineer in Salford and experiments in 1905 showed that his method was superior in that greater volumes of sewage could be purified better for longer periods of time than could be achieved by the contact bed.

The Royal Commission on Sewage Disposal published its eighth report in 1912 that set what became the international standard for sewage discharge into rivers; the '20:30 standard', which allowed 20 mg Biochemical oxygen demand and 30 mg suspended solid per litre.

Biofilter

Biosolids composting plant biofilter mound - note sprinkler visible front right to maintain proper moisture level for optimum functioning

Biofiltration is a pollution control technique using living material to capture and biologically degrade pollutants. Common uses include processing waste water, capturing harmful chemicals or silt from surface runoff, and microbiotic oxidation of contaminants in air.

Examples of biofiltration include;

- Bioswales, biostrips, biobags, bioscrubbers, and trickling filters

- Constructed wetlands and natural wetlands

- Slow sand filters

- Treatment ponds

- Green belts

- Green walls

- Riparian zones, riparian forests, bosques

Control of Air Pollution

When applied to air filtration and purification, biofilters use microorganisms to remove air pollution. The air flows through a packed bed and the pollutant transfers into a thin biofilm on the surface of the packing material. Microorganisms, including bacteria and fungi are immobilized in the biofilm and degrade the pollutant. Trickling filters and bioscrubbers rely on a biofilm and the bacterial action in their recirculating waters.

The technology finds greatest application in treating malodorous compounds and water-soluble volatile organic compounds (VOCs). Industries employing the technology include food and animal products, off-gas from wastewater treatment facilities, pharmaceuticals, wood products manufacturing, paint and coatings application and manufacturing and resin manufacturing and application, etc. Compounds treated are typically mixed VOCs and various sulfur compounds, including hydrogen sulfide. Very large airflows may be treated and although a large area (footprint) has typically been required—a large biofilter (>200,000 acfm) may occupy as much or more land than a football field—this has been one of the principal drawbacks of the technology. Engineered biofilters, designed and built since the early 1990s, have provided significant footprint reductions over the conventional flat-bed, organic media type.

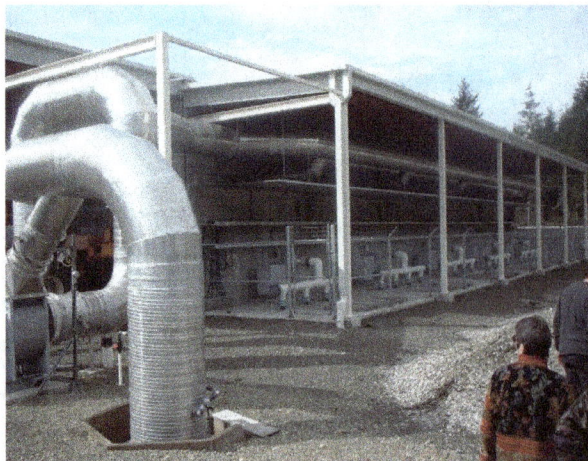

Air cycle system at biosolids composting plant. Large duct in foreground is exhaust air into biofilter shown in next photo

One of the main challenges to optimum biofilter operation is maintaining proper moisture throughout the system. The air is normally humidified before it enters the bed with a watering (spray) system, humidification chamber, bioscrubber, or biotrickling filter. Properly maintained, a natural, organic packing media like peat, vegetable mulch, bark or wood chips may last for several years but engineered, combined natural organic and synthetic component packing materials will generally last much longer, up to 10 years. A number of companies offer these types or proprietary packing materials and multi-year guarantees, not usually provided with a conventional compost or wood chip bed biofilter.

Although widely employed, the scientific community is still unsure of the physical phenomena underpinning biofilter operation, and information about the microorganisms involved continues to be developed. A biofilter/bio-oxidation system is a fairly simple device to construct and operate and offers a cost-effective solution provided the pollutant is biodegradable within a moderate time frame (increasing residence time = increased size and capital costs), at reasonable concentrations (and lb/hr loading rates) and that the airstream is at an organism-viable temperature. For large volumes of air, a biofilter may be the only cost-effective solution. There is no secondary pollution (unlike the case of incineration where additional CO_2 and NO_x are produced from burning fuels) and degradation products form additional biomass, carbon dioxide and water. Media irrigation water, although many systems recycle part of it to reduce operating costs, has a moderately high biochemical oxygen demand (BOD) and may require treatment before disposal. However, this "blowdown water", necessary for proper maintenance of any bio-oxidation system, is generally accepted by municipal publicly owned treatment works without any pretreatment.

Biofilters are being utilized in Columbia Falls, Montana at Plum Creek Timber Company's fiberboard plant. The biofilters decrease the pollution emitted by the manufacturing process and the exhaust emitted is 98% clean. The newest, and largest, biofilter addition to Plum Creek cost $9.5 million, yet even though this new technology is expensive, in the long run it will cost less overtime than the alternative exhaust-cleaning incinerators fueled by natural gas (which are not as environmentally friendly). The biofilters use trillions of microscopic bacteria that cleanse the air being released from the plant.

Water Treatment

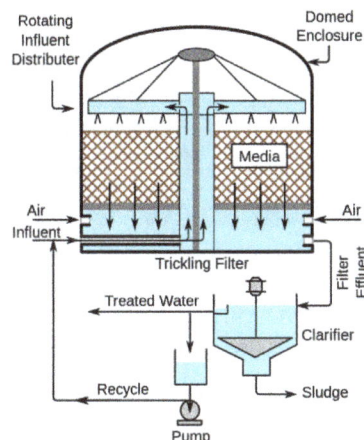

A typical complete trickling filter system for treating wastewaters.

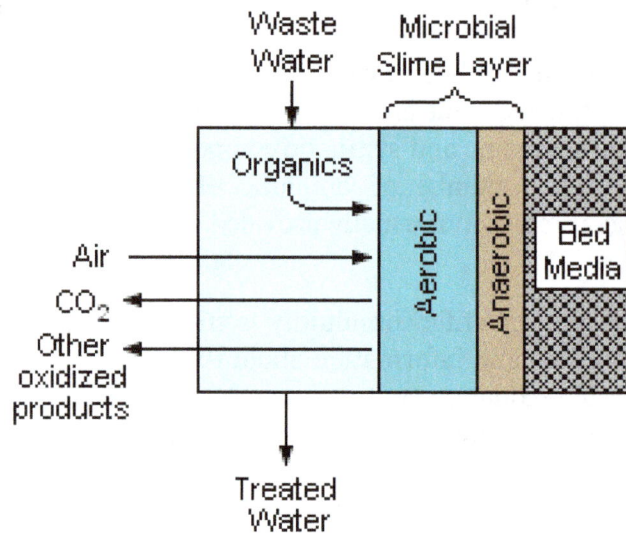

Image 1: A schematic cross-section of the contact face of the bed media in a trickling filter.

Biofiltration Process

Biofilter installation at a commercial composting facility.

A biofilter is a bed of media on which microorganisms attach and grow to form a biological layer called biofilm. Biofiltration is thus usually referred to as a fixed–film process. Generally, the biofilm is formed by a community of different microorganisms (bacteria, fungi, yeast, etc.), macro-organisms (protozoa, worms, insect's larvae, etc.) and extracellular polymeric substances (EPS) (Flemming and Wingender, 2010). The aspect of the biofilm is usually slimy and muddy.

Water to be treated can be applied intermittently or continuously over the media, via upflow or downflow. Typically, a biofilter has two or three phases, depending on the feeding strategy (percolating or submerged biofilter):

- a solid phase (media);
- a liquid phase (water);
- a gaseous phase (air).

Organic matter and other water components diffuse into the biofilm where the treatment occurs, mostly by biodegradation. Biofiltration processes are usually aerobic, which means that microorganisms require oxygen for their metabolism. Oxygen can be supplied to the biofilm, either concurrently or countercurrently with water flow. Aeration occurs passively by the natural flow of air through the process (three phase biofilter) or by forced air supplied by blowers.

Microorganisms' activity is a key-factor of the process performance. The main influencing factors are the water composition, the biofilter hydraulic loading, the type of media, the feeding strategy (percolation or submerged media), the age of the biofilm, temperature, aeration, etc.

Types of Filtering Media

Originally, biofilter was developed using rock or slag as filter media, but different types of material are used today. These materials are categorized as inorganic media (sand, gravel, geotextile, different shapes of plastic media, glass beads, etc.) and organic media (peat, wood chips, coconut shell fragments, compost, etc.)

Advantages

Although biological filters have simple superficial structures, their internal hydrodynamics and the microorganisms' biology and ecology are complex and variable. These characteristics confer robustness to the process. In other words, the process has the capacity to maintain its performance or rapidly return to initial levels following a period of no flow, of intense use, toxic shocks, media backwash (high rate biofiltration processes), etc.

The structure of the biofilm protects microorganisms from difficult environmental conditions and retains the biomass inside the process, even when conditions are not optimal for its growth. Biofiltration processes offer the following advantages: (Rittmann et al., 1988):

- Because microorganisms are retained within the biofilm, biofiltration allows the development of microorganisms with relatively low specific growth rates;
- Biofilters are less subject to variable or intermittent loading and to hydraulic shock;
- Operational costs are usually lower than for activated sludge;
- Final treatment result is less influenced by biomass separation since the biomass concentration at the effluent is much lower than for suspended biomass processes;
- Attached biomass becomes more specialized (higher concentration of relevant organisms) at a given point in the process train because there is no biomass return.

Drawbacks

Because filtration and growth of biomass leads to an accumulation of matter in the filtering media, this type of fixed-film process is subject to clogging and flow channeling. Depending on the type of application and on the media used for microbial growth, clogging can be controlled using physical and/or chemical methods. Whenever possible, backwash steps can be implemented using air and/or water to disrupt the biomat and recover flow. Chemicals such as oxidizing (peroxide, ozone) or biocide agents can also be used.

Drinking Water

For drinking water, biological water treatment involves the use of naturally-occurring microorganisms in the surface water to improve water quality. Under optimum conditions, including relatively low turbidity and high oxygen content, the organisms break down material in the water and thus improve water quality. Slow sand filters or carbon filters are used to provide a support on which these microorganisms grow. These biological treatment systems effectively reduce water-borne diseases, dissolved organic carbon, turbidity and color in surface water, thus improving overall water quality.

Wastewater

Biofiltration is used to treat wastewater from a wide range of sources, with varying organic compositions and concentrations. Many examples of biofiltration applications are described in the literature. As a non-exhaustive list of applications, and notwithstanding the type of media, biofilters were developed and commercialized for the treatment of animal wastes, landfill leachates, dairy wastewater, domestic wastewater.

This process is versatile as it can be adapted to small flows (< 1 m3/d), such as onsite domestic wastewater as well as to flows generated by a municipality (> 240 000 m3/d). For decentralized domestic wastewater production, such as for isolated dwellings, it has been demonstrated that there are important daily, weekly and yearly fluctuations of hydraulic and organic production rates related to modern families' lifestyle. In this context, a biofilter located after a septic tank constitutes a robust process able to sustain the variability observed without compromising the treatment performance.

Use in Aquaculture

The use of biofilters is common in closed aquaculture systems, such as recirculating aquaculture systems (RAS). Many designs are used, with different benefits and drawbacks, however the function is the same: reducing water exchanges by converting ammonia to nitrate. Ammonia (NH_4^+ and NH_3) originates from the brachial excretion from the gills of aquatic animals and from the decomposition of organic matter. As ammonia-N is highly toxic, this is converted to a less toxic form of nitrite (by *Nitrosomonas* sp.) and then to an even less toxic form of nitrate (by *Nitrobacter* sp.). This "nitrification" process requires oxygen (aerobic conditions), without which the biofilter can crash. Furthermore, as this nitrification cycle produces H^+, the pH can decrease which necessitates the use of buffers such as lime.

Industrial Wastewater Treatment

Industrial wastewater treatment covers the mechanisms and processes used to treat wastewater that is produced as a by-product of industrial or commercial activities. After treatment, the treated industrial wastewater (or effluent) may be reused or released to a sanitary sewer or to a surface water in the environment. Most industries produce some wastewater although recent trends in the developed world have been to minimise such production or recycle such wastewater within the production process. However, many industries remain dependent on processes that produce wastewaters.

Sources of Industrial Wastewater

Complex Organic Chemicals Industry

A range of industries manufacture or use complex organic chemicals. These include pesticides, pharmaceuticals, paints and dyes, petrochemicals, detergents, plastics, paper pollution, etc. Waste waters can be contaminated by feedstock materials, by-products, product material in soluble or particulate form, washing and cleaning agents, solvents and added value products such as plasticisers. Treatment facilities that do not need control of their effluent typically opt for a type of aerobic treatment, i.e. aerated lagoons.

Electric Power Plants

Fossil-fuel power stations, particularly coal-fired plants, are a major source of industrial wastewater. Many of these plants discharge wastewater with significant levels of metals such as lead, mercury, cadmium and chromium, as well as arsenic, selenium and nitrogen compounds (nitrates and nitrites). Wastewater streams include flue-gas desulfurization, fly ash, bottom ash and flue gas mercury control. Plants with air pollution controls such as wet scrubbers typically transfer the captured pollutants to the wastewater stream.

Ash ponds, a type of surface impoundment, are a widely used treatment technology at coal-fired plants. These ponds use gravity to settle out large particulates (measured as total suspended solids) from power plant wastewater. This technology does not treat dissolved pollutants. Power stations use additional technologies to control pollutants, depending on the particular wastestream in the plant. These include dry ash handling, closed-loop ash recycling, chemical precipitation, biological treatment (such as an activated sludge process), and evaporation.

Food Industry

Wastewater generated from agricultural and food operations has distinctive characteristics that set it apart from common municipal wastewater managed by public or private sewage treatment plants throughout the world: it is biodegradable and non-toxic, but has high concentrations of biochemical oxygen demand (BOD) and suspended solids (SS). The constituents of food and agriculture wastewater are often complex to predict, due to the differences in BOD and pH in effluents from vegetable, fruit, and meat products and due to the seasonal nature of food processing and post-harvesting.

Processing of food from raw materials requires large volumes of high grade water. Vegetable washing generates waters with high loads of particulate matter and some dissolved organic matter. It may also contain surfactants.

Animal slaughter and processing produces very strong organic waste from body fluids, such as blood, and gut contents. This wastewater is frequently contaminated by significant levels of antibiotics and growth hormones from the animals and by a variety of pesticides used to control external parasites.

Processing food for sale produces wastes generated from cooking which are often rich in plant organic material and may also contain salt, flavourings, colouring material and acids or alkali. Very significant quantities of oil or fats may also be present.

Iron and Steel Industry

The production of iron from its ores involves powerful reduction reactions in blast furnaces. Cooling waters are inevitably contaminated with products especially ammonia and cyanide. Production of coke from coal in coking plants also requires water cooling and the use of water in by-products separation. Contamination of waste streams includes gasification products such as benzene, naphthalene, anthracene, cyanide, ammonia, phenols, cresols together with a range of more complex organic compounds known collectively as polycyclic aromatic hydrocarbons (PAH).

The conversion of iron or steel into sheet, wire or rods requires hot and cold mechanical transformation stages frequently employing water as a lubricant and coolant. Contaminants include hydraulic oils, tallow and particulate solids. Final treatment of iron and steel products before onward sale into manufacturing includes *pickling* in strong mineral acid to remove rust and prepare the surface for tin or chromium plating or for other surface treatments such as galvanisation or painting. The two acids commonly used are hydrochloric acid and sulfuric acid. Wastewaters include acidic rinse waters together with waste acid. Although many plants operate acid recovery plants (particularly those using hydrochloric acid), where the mineral acid is boiled away from the iron salts, there remains a large volume of highly acid ferrous sulfate or ferrous chloride to be disposed of. Many steel industry wastewaters are contaminated by hydraulic oil, also known as *soluble oil*.

Mines and Quarries

Mine wastewater effluent in Peru, with neutralized pH from tailing runoff.

The principal waste-waters associated with mines and quarries are slurries of rock particles in water. These arise from rainfall washing exposed surfaces and haul roads and also from rock washing and grading processes. Volumes of water can be very high, especially rainfall related arisings on large sites. Some specialized separation operations, such as coal washing to separate coal from native rock using density gradients, can produce wastewater contaminated by fine particulate haematite and surfactants. Oils and hydraulic oils are also common contaminants.

Wastewater from metal mines and ore recovery plants are inevitably contaminated by the minerals present in the native rock formations. Following crushing and extraction of the desirable materials, undesirable materials may enter the wastewater stream. For metal mines, this can include unwanted metals such as zinc and other materials such as arsenic. Extraction of high value metals such as gold and silver may generate slimes containing very fine particles in where physical removal of contaminants becomes particularly difficult.

Additionally, the geologic formations that harbour economically valuable metals such as copper and gold very often consist of sulphide-type ores. The processing entails grinding the rock into fine particles and then extracting the desired metal(s), with the leftover rock being known as tailings. These tailings contain a combination of not only undesirable leftover metals, but also sulphide components which eventually form sulphuric acid upon the exposure to air and water that inevitably occurs when the tailings are disposed of in large impoundments. The resulting acid mine drainage, which is often rich in heavy metals (because acids dissolve metals), is one of the many environmental impacts of mining.

Nuclear Industry

The waste production from the nuclear and radio-chemicals industry is dealt with as *Radioactive waste*.

Pulp and Paper Industry

Effluent from the pulp and paper industry is generally high in suspended solids and BOD. Plants that bleach wood pulp for paper making may generate chloroform, dioxins (including 2,3,7,8-TCDD), furans, phenols and chemical oxygen demand (COD). Stand-alone paper mills using imported pulp may only require simple primary treatment, such as sedimentation or dissolved air flotation. Increased BOD or COD loadings, as well as organic pollutants, may require biological treatment such as activated sludge or upflow anaerobic sludge blanket reactors. For mills with high inorganic loadings like salt, tertiary treatments may be required, either general membrane treatments like ultrafiltration or reverse osmosis or treatments to remove specific contaminants, such as nutrients.

Industrial Oil Contamination

Industrial applications where oil enters the wastewater stream may include vehicle wash bays, workshops, fuel storage depots, transport hubs and power generation. Often the wastewater is discharged into local sewer or trade waste systems and must meet local environmental specifications. Typical contaminants can include solvents, detergents, grit. lubricants and hydrocarbons.

Water Treatment

Many industries have a need to treat water to obtain very high quality water for demanding purposes such as environmental discharge compliance. Water treatment produces organic and mineral sludges from filtration and sedimentation. Ion exchange using natural or synthetic resins removes calcium, magnesium and carbonate ions from water, typically replacing them with sodium, chloride, hydroxyl and/or other ions. Regeneration of ion exchange columns with strong acids and alkalis produces a wastewater rich in hardness ions which are readily precipitated out, especially when in admixture with other wastewater constituents.

Wool Processing

Insecticide residues in fleeces are a particular problem in treating waters generated in wool processing. Animal fats may be present in the wastewater, which if not contaminated, can be recovered for the production of tallow or further rendering.

Treatment of Industrial Wastewater

The various types of contamination of wastewater require a variety of strategies to remove the contamination.

Brine Treatment

Brine treatment involves removing dissolved salt ions from the waste stream. Although similarities to seawater or brackish water desalination exist, industrial brine treatment may contain unique combinations of dissolved ions, such as hardness ions or other metals, necessitating specific processes and equipment.

Brine treatment systems are typically optimized to either reduce the volume of the final discharge for more economic disposal (as disposal costs are often based on volume) or maximize the recovery of fresh water or salts. Brine treatment systems may also be optimized to reduce electricity consumption, chemical usage, or physical footprint.

Brine treatment is commonly encountered when treating cooling tower blowdown, produced water from steam assisted gravity drainage (SAGD), produced water from natural gas extraction such as coal seam gas, frac flowback water, acid mine or acid rock drainage, reverse osmosis reject, chlor-alkali wastewater, pulp and paper mill effluent, and waste streams from food and beverage processing.

Brine treatment technologies may include: membrane filtration processes, such as reverse osmosis; ion exchange processes such as electrodialysis or weak acid cation exchange; or evaporation processes, such as brine concentrators and crystallizers employing mechanical vapour recompression and steam.

Reverse osmosis may not be viable for brine treatment, due to the potential for fouling caused by hardness salts or organic contaminants, or damage to the reverse osmosis membranes from hydrocarbons.

Evaporation processes are the most widespread for brine treatment as they enable the highest degree of concentration, as high as solid salt. They also produce the highest purity effluent, even distillate-quality. Evaporation processes are also more tolerant of organics, hydrocarbons, or hardness salts. However, energy consumption is high and corrosion may be an issue as the prime mover is concentrated salt water. As a result, evaporation systems typically employ titanium or duplex stainless steel materials.

Brine Management

Brine management examines the broader context of brine treatment and may include consideration of government policy and regulations, corporate sustainability, environmental impact, recycling, handling and transport, containment, centralized compared to on-site treatment, avoidance and reduction, technologies, and economics. Brine management shares some issues with leachate management and more general waste management.

Solids Removal

Most solids can be removed using simple sedimentation techniques with the solids recovered as slurry or sludge. Very fine solids and solids with densities close to the density of water pose special problems. In such case filtration or ultrafiltration may be required. Although, flocculation may be used, using alum salts or the addition of polyelectrolytes.

Oils and Grease Removal

The effective removal of oils and grease is dependent on the characteristics of the oil in terms of its suspension state and droplet size, which will in turn affect the choice of separator technology.

Oil pollution in water usually comes in four states, often in combination:

- free oil - large oil droplets sitting on the surface;
- heavy oil, which sits at the bottom, often adhering to solids like dirt;
- emulsified, where the oil droplets are heavily "chopped"; and
- dissolved oil, where the droplets are fully dispersed and not visible. Emulsified oil droplets are the most common in industrial oily wastewater and are extremely difficult to separate.

The methodology for separating the oil is dependent on the oil droplet size. Larger oil droplets such as those in free oil pollution are easily removed, but as the droplets become smaller, some separator technologies perform better than others.

Most separator technologies will have an optimum range of oil droplet sizes that can be effectively treated. This is known as the "micron rating."

Analysing the oily water to determine droplet size can be performed with a video particle analyser. Alternatively, there are commonalities in industries for oil droplet sizes. Larger droplets–greater than 60 microns–are often present in wastewater in workshops, re-fuel areas and depots. Twenty to 50 micron oil droplets often are present in vehicle wash bays, meat processing and dairy manufacturing effluent and aluminium billet cooling towers. Smaller droplets in the range of 10 to 20 microns tend to occur in workshops and condensates.

Each separator technology will have its' own performance curve outlining optimum performance based on oil droplet size. the most common separators are gravity tanks or pits, API oil-water separators or plate packs, chemical treatment via DAFs, centrifuges, media filters and hydrocyclones.

API separators

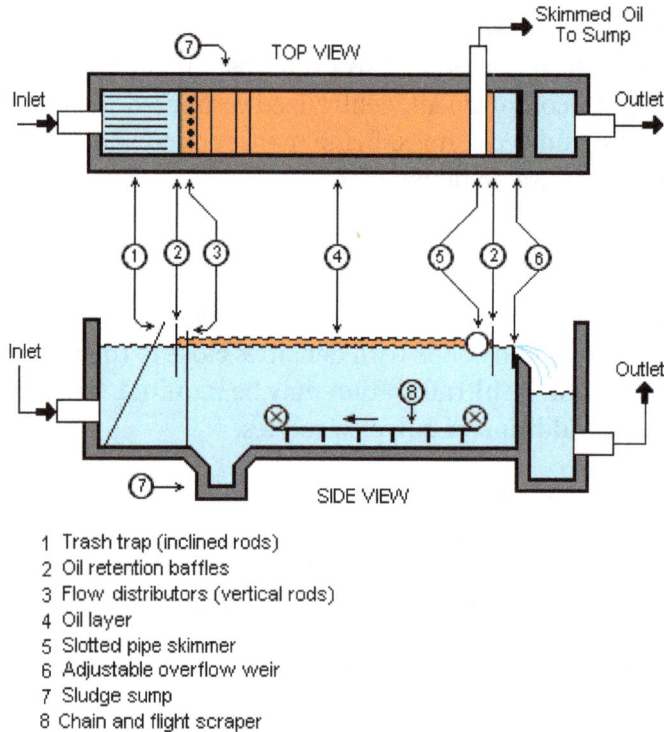

1 Trash trap (inclined rods)
2 Oil retention baffles
3 Flow distributors (vertical rods)
4 Oil layer
5 Slotted pipe skimmer
6 Adjustable overflow weir
7 Sludge sump
8 Chain and flight scraper

A typical API oil-water separator used in many industries

Many oils can be recovered from open water surfaces by skimming devices. Considered a dependable and cheap way to remove oil, grease and other hydrocarbons from water, oil skimmers can sometimes achieve the desired level of water purity. At other times, skimming is also a cost-efficient method to remove most of the oil before using membrane filters and chemical processes. Skimmers will prevent filters from blinding prematurely and keep chemical costs down because there is less oil to process.

Because grease skimming involves higher viscosity hydrocarbons, skimmers must be equipped with heaters powerful enough to keep grease fluid for discharge. If floating grease forms into solid clumps or mats, a spray bar, aerator or mechanical apparatus can be used to facilitate removal.

However, hydraulic oils and the majority of oils that have degraded to any extent will also have a soluble or emulsified component that will require further treatment to eliminate. Dissolving or emulsifying oil using surfactants or solvents usually exacerbates the problem rather than solving it, producing wastewater that is more difficult to treat.

The wastewaters from large-scale industries such as oil refineries, petrochemical plants, chemical plants, and natural gas processing plants commonly contain gross amounts of oil and suspended solids. Those industries use a device known as an API oil-water separator which is designed to separate the oil and suspended solids from their wastewater effluents. The name is derived from the fact that such separators are designed according to standards published by the American Petroleum Institute (API).

The API separator is a gravity separation device designed by using Stokes Law to define the rise velocity of oil droplets based on their density and size. The design is based on the specific gravity difference between the oil and the wastewater because that difference is much smaller than the specific gravity difference between the suspended solids and water. The suspended solids settles to the bottom of the separator as a sediment layer, the oil rises to top of the separator and the cleansed wastewater is the middle layer between the oil layer and the solids.

Typically, the oil layer is skimmed off and subsequently re-processed or disposed of, and the bottom sediment layer is removed by a chain and flight scraper (or similar device) and a sludge pump. The water layer is sent to further treatment for additional removal of any residual oil and then to some type of biological treatment unit for removal of undesirable dissolved chemical compounds.

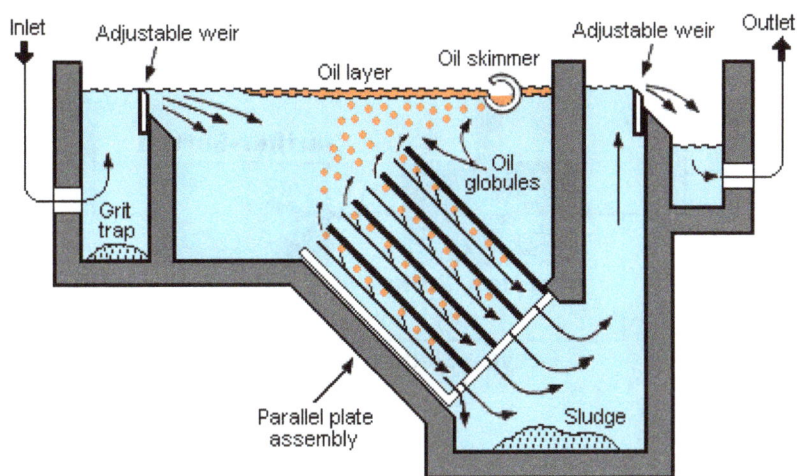

A typical parallel plate separator

Parallel plate separators are similar to API separators but they include tilted parallel plate assemblies (also known as parallel packs). The parallel plates provide more surface for suspended oil droplets to coalesce into larger globules. Such separators still depend upon the specific gravity between the suspended oil and the water. However, the parallel plates enhance the degree of oil-water separation. The result is that a parallel plate separator requires significantly less space than a conventional API separator to achieve the same degree of separation.

Hydrocyclone Oil Separators

Hydrocyclone oil separators operate on the process where wastewater enters the cyclone chamber and is spun under extreme centrifugal forces more than 1000 times the force of gravity. This force causes the water and oil droplets to separate. The separated oil is discharged from one end of the cyclone where treated water is discharged through the opposite end for further treatment, filtration or discharge.

Hydrocyclones are useful for the greatest range of oil droplet sizes operating from less than 10 microns and up and can operate continuously without water pre-treatment and at any temperature and pH. Applications where hydrocyclones are found are in industry where oily water sources arise in workshops, vehicle wash bays, transport hubs, fuel depots and aluminium billet processing. Animal fats from meat processing and dairy manufacturing can also be removed without the need of chemical treatment that often is required for dissolved air flotation (DAF) systems.

Removal of Biodegradable Organics

Biodegradable organic material of plant or animal origin is usually possible to treat using extended conventional sewage treatment processes such as activated sludge or trickling filter. Problems can arise if the wastewater is excessively diluted with washing water or is highly concentrated such as undiluted blood or milk. The presence of cleaning agents, disinfectants, pesticides, or antibiotics can have detrimental impacts on treatment processes.

Activated Sludge Process

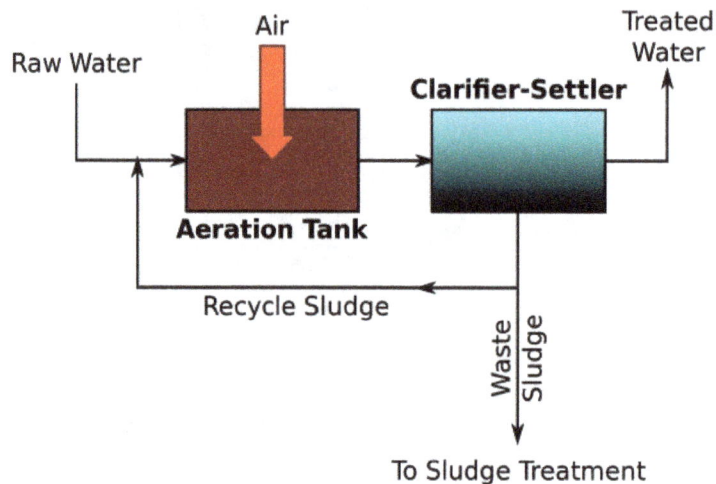

A generalized diagram of an activated sludge process.

Activated sludge is a biochemical process for treating sewage and industrial wastewater that uses air (or oxygen) and microorganisms to biologically oxidize organic pollutants, producing a waste sludge (or floc) containing the oxidized material. In general, an activated sludge process includes:

- An aeration tank where air (or oxygen) is injected and thoroughly mixed into the wastewater.

- A settling tank (usually referred to as a clarifier or "settler") to allow the waste sludge to settle. Part of the waste sludge is recycled to the aeration tank and the remaining waste sludge is removed for further treatment and ultimate disposal.

Trickling Filter Process

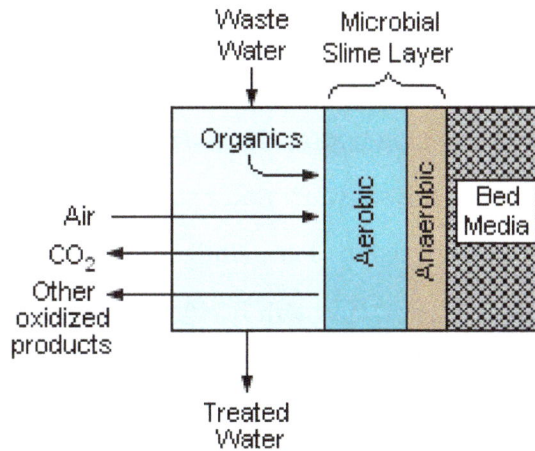

A schematic cross-section of the contact face of the bed media in a trickling filter

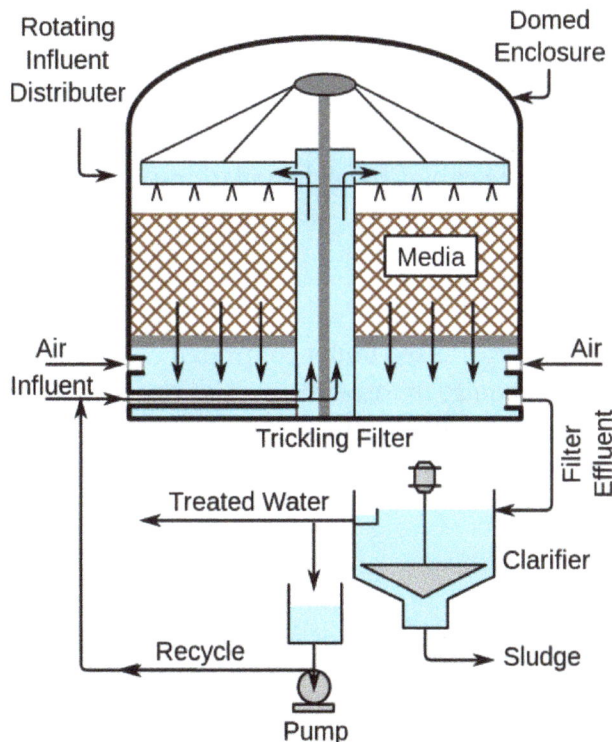

A typical complete trickling filter system

A trickling filter consists of a bed of rocks, gravel, slag, peat moss, or plastic media over which wastewater flows downward and contacts a layer (or film) of microbial slime covering the bed media. Aerobic conditions are maintained by forced air flowing through the bed or by natural convection of air. The process involves adsorption of organic compounds in the wastewater by the microbial slime layer, diffusion of air into the slime layer to provide the oxygen required for the biochemical oxidation of the organic compounds. The end products include carbon dioxide gas, water and other products of the oxidation. As the slime layer thickens, it becomes difficult for the air to penetrate the layer and an inner anaerobic layer is formed.

The fundamental components of a complete trickling filter system are:

- A bed of filter medium upon which a layer of microbial slime is promoted and developed.

- An enclosure or a container which houses the bed of filter medium.

- A system for distributing the flow of wastewater over the filter medium.

- A system for removing and disposing of any sludge from the treated effluent.

The treatment of sewage or other wastewater with trickling filters is among the oldest and most well characterized treatment technologies.

A trickling filter is also often called a *trickle filter*, *trickling biofilter*, *biofilter*, *biological filter* or *biological trickling filter*.

Treatment of Other Organics

Synthetic organic materials including solvents, paints, pharmaceuticals, pesticides, products from coke production and so forth can be very difficult to treat. Treatment methods are often specific to the material being treated. Methods include advanced oxidation processing, distillation, adsorption, vitrification, incineration, chemical immobilisation or landfill disposal. Some materials such as some detergents may be capable of biological degradation and in such cases, a modified form of wastewater treatment can be used.

Treatment of Acids and Alkalis

Acids and alkalis can usually be neutralised under controlled conditions. Neutralisation frequently produces a precipitate that will require treatment as a solid residue that may also be toxic. In some cases, gases may be evolved requiring treatment for the gas stream. Some other forms of treatment are usually required following neutralisation.

Waste streams rich in hardness ions as from de-ionisation processes can readily lose the hardness ions in a buildup of precipitated calcium and magnesium salts. This precipitation process can cause severe *furring* of pipes and can, in extreme cases, cause the blockage of disposal pipes. A 1 metre diameter industrial marine discharge pipe serving a major chemicals complex was blocked by such salts in the 1970s. Treatment is by concentration of de-ionisation waste waters and disposal to landfill or by careful pH management of the released wastewater.

Treatment of Toxic Materials

Toxic materials including many organic materials, metals (such as zinc, silver, cadmium, thallium, etc.) acids, alkalis, non-metallic elements (such as arsenic or selenium) are generally resistant to biological processes unless very dilute. Metals can often be precipitated out by changing the pH or by treatment with other chemicals. Many, however, are resistant to treatment or mitigation and may require concentration followed by landfilling or recycling. Dissolved organics can be *incinerated* within the wastewater by the advanced oxidation process.

Phytoremediation

Phytoremediation refers to the technologies that use living plants to clean up soil, air, and water contaminated with hazardous chemicals.

Phytoremediation is a cost-effective plant-based approach of remediation that takes advantage of the ability of plants to concentrate elements and compounds from the environment and to metabolize various molecules in their tissues. It refers to the natural ability of certain plants called hyperaccumulators to bioaccumulate, degrade, or render harmless contaminants in soils, water, or air. Toxic heavy metals and organic pollutants are the major targets for phytoremediation. Knowledge of the physiological and molecular mechanisms of phytoremediation began to emerge in recent years together with biological and engineering strategies designed to optimize and improve phytoremediation. In addition, several field trials confirmed the feasibility of using plants for environmental cleanup.

Application

Phytoremediation may be applied wherever the soil or static water environment has become polluted or is suffering ongoing chronic pollution. Examples where phytoremediation has been used successfully include the restoration of abandoned metal mine workings, and sites where polychlorinated biphenyls have been dumped during manufacture and mitigation of ongoing coal mine discharges reducing the impact of contaminants in soils, water, or air. Contaminants such as metals, pesticides, solvents, explosives, and crude oil and its derivatives, have been mitigated in phytoremediation projects worldwide. Many plants such as mustard plants, alpine pennycress, hemp, and pigweed have proven to be successful at hyperaccumulating contaminants at toxic waste sites.

Over the past 20 years, this technology has become increasingly popular and has been employed at sites with soils contaminated with lead, uranium, and arsenic. While it has the advantage that environmental concerns may be treated in situ; one major disadvantage of phytoremediation is that it requires a long-term commitment, as the process is dependent on a plant's ability to grow and thrive in an environment that is not ideal for normal plant growth.

Advantages and Limitations

- Advantages:
  o the cost of the phytoremediation is lower than that of traditional processes both *in situ* and *ex situ*

  o the plants can be easily monitored

  o the possibility of the recovery and re-use of valuable metals (by companies specializing in "phyto mining")

  o it is potentially the least harmful method because it uses naturally occurring organisms and preserves the environment in a more natural state.

- Limitations:

  o phytoremediation is limited to the surface area and depth occupied by the roots.

  o slow growth and low biomass require a long-term commitment

  o with plant-based systems of remediation, it is not possible to completely prevent the leaching of contaminants into the groundwater (without the complete removal of the contaminated ground, which in itself does not resolve the problem of contamination)

  o the survival of the plants is affected by the toxicity of the contaminated land and the general condition of the soil.

  o bio-accumulation of contaminants, especially metals, into the plants which then pass into the food chain, from primary level consumers upwards or requires the safe disposal of the affected plant material.

Various Phytoremediation Processes

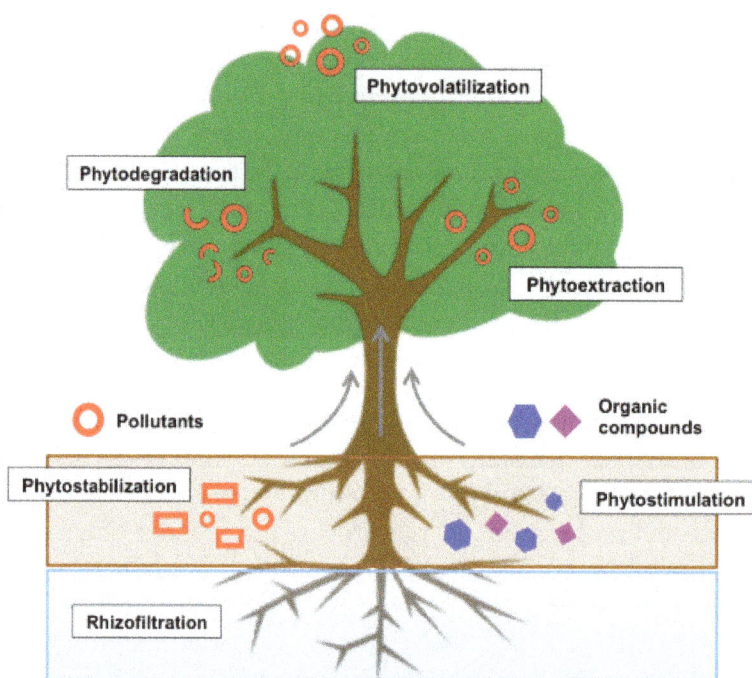

Phytoremediation process

A range of processes mediated by plants or algae are useful in treating environmental problems:

- *Phytosequestration* – Phytochemical complexation in the root zone, reduce the fraction of the contaminant that is bioavailable. Transport protein inhibition on the root membrane-preventing contaminants from entering the plant. Vacuolar storage in the root cells: contaminants can be sequestered into the vacuoles of root cells.

- *Phytoextraction* – uptake and concentration of substances from the environment into the plant biomass.

- *Phytostabilization* – reducing the mobility of substances in the environment, for example, by limiting the leaching of substances from the soil.

- *Phytotransformation* – chemical modification of environmental substances as a direct result of plant metabolism, often resulting in their inactivation, degradation (phytodegradation), or immobilization (phytostabilization).

- *Phytostimulation* – enhancement of soil microbial activity for the degradation of contaminants, typically by organisms that associate with roots. This process is also known as *rhizosphere degradation*. Phytostimulation can also involve aquatic plants supporting active populations of microbial degraders, as in the stimulation of atrazine degradation by hornwort.

- *Phytovolatilization* – removal of substances from soil or water with release into the air, sometimes as a result of phytotransformation to more volatile and/or less polluting substances.

- *Rhizofiltration* – filtering water through a mass of roots to remove toxic substances or excess nutrients. The pollutants remain absorbed in or adsorbed to the roots.

- Biological hydraulic containment – Some plants, like poplars, draws water upwards through the soil into the roots and out through the plant decreases the movement of soluble contaminants downwards, deeper into the site and into the groundwater.

Phytoextraction

Phytoextraction (or *phytoaccumulation*) uses plants or algae to remove contaminants from soils, sediments or water into harvestable plant biomass (organisms that take larger-than-normal amounts of contaminants from the soil are called hyperaccumulators). Phytoextraction has been growing rapidly in popularity worldwide for the last twenty years or so. In general, this process has been tried more often for extracting heavy metals than for organics. At the time of disposal, contaminants are typically concentrated in the much smaller volume of the plant matter than in the initially contaminated soil or sediment. 'Mining with plants', or phytomining, is also being experimented with:

The plants absorb contaminants through the root system and store them in the root biomass and/or transport them up into the stems and/or leaves. A living plant may continue to absorb contaminants until it is harvested. After harvest, a lower level of the contaminant will remain in the soil, so the growth/harvest cycle must usually be repeated through several crops to achieve a significant cleanup. After the process, the cleaned soil can support other vegetation.

Advantages

The main advantage of phytoextraction is environmental friendliness. Traditional methods that are used for cleaning up heavy metal-contaminated soil disrupt soil structure and reduce soil productivity, whereas phytoextraction can clean up the soil without causing any kind of harm to soil quality. Another benefit of phytoextraction is that it is less expensive than any other clean-up process.

Disadvantages

As this process is controlled by plants, it takes more time than anthropogenic soil clean-up methods.

Two Versions of Phytoextraction

- natural hyper-accumulation, where plants naturally take up the contaminants in soil unassisted.

- induced or assisted hyper-accumulation, where a conditioning fluid containing a chelator or another agent is added to soil to increase metal solubility or mobilization so that the plants can absorb them more easily. In many cases natural hyperaccumulators are metallophyte plants that can tolerate and incorporate high levels of toxic metals.

Examples of Phytoextraction

- Arsenic, using the sunflower (*Helianthus annuus*), or the Chinese Brake fern (*Pteris vittata*).

- Cadmium, using willow (*Salix viminalis*): In 1999, one research experiment performed by Maria Greger and Tommy Landberg suggested willow has a significant potential as a phytoextractor of cadmium (Cd), zinc (Zn), and copper (Cu), as willow has some specific characteristics like high transport capacity of heavy metals from root to shoot and huge amount of biomass production; can be used also for production of bio energy in the biomass energy power plant.

- Cadmium and zinc, using alpine pennycress (*Thlaspi caerulescens*), a hyperaccumulator of these metals at levels that would be toxic to many plants. On the other hand, the presence of copper seems to impair its growth.

- Lead, using Indian mustard (*Brassica juncea*), ragweed (*Ambrosia artemisiifolia*), hemp dogbane (*Apocynum cannabinum*), or poplar trees, which sequester lead in their biomass.

- Salt-tolerant (moderately halophytic) barley and/or sugar beets are commonly used for the extraction of sodium chloride (common salt) to reclaim fields that were previously flooded by sea water.

- Caesium-137 and strontium-90 were removed from a pond using sunflowers after the Chernobyl accident.

- Mercury, selenium and organic pollutants such as polychlorinated biphenyls (PCBs) have been removed from soils by transgenic plants containing genes for bacterial enzymes.

Phytostabilization

Phytostabilization focuses on the long term stabilization and containment of the pollutant.For example, the plant's presence can reduce wind erosion; or the plant's roots can prevent water erosion, immobilize the pollutants by adsorption or accumulation, and provide a zone around the roots where the pollutant can precipitate and stabilize. Unlike phytoextraction, phytostabilization focuses mainly on sequestering pollutants in soil near the roots but not in plant tissues. Pollutants become less bioavailable, and livestock, wildlife, and human exposure is reduced. An example application of this sort is using a vegetative cap to stabilize and contain mine tailings.

Phytotransformation

In the case of organic pollutants, such as pesticides, explosives, solvents, industrial chemicals, and other xenobiotic substances, certain plants, such as Cannas, render these substances non-toxic by their metabolism. In other cases, microorganisms living in association with plant roots may metabolize these substances in soil or water. These complex and recalcitrant compounds cannot be broken down to basic molecules (water, carbon-dioxide, etc.) by plant molecules, and, hence, the term *phytotransformation* represents a change in chemical structure without complete breakdown of the compound. The term "Green Liver" is used to describe phytotransformation, as plants behave analogously to the human liver when dealing with these xenobiotic compounds (foreign compound/pollutant). After uptake of the xenobiotics, plant enzymes increase the polarity of the xenobiotics by adding functional groups such as hydroxyl groups (-OH).

This is known as Phase I metabolism, similar to the way that the human liver increases the polarity of drugs and foreign compounds (drug metabolism). Whereas in the human liver enzymes such as cytochrome P450s are responsible for the initial reactions, in plants enzymes such as peroxidases, phenoloxidases, esterases and nitroreductases carry out the same role.

In the second stage of phytotransformation, known as Phase II metabolism, plant biomolecules such as glucose and amino acids are added to the polarized xenobiotic to further increase the polarity (known as conjugation). This is again similar to the processes occurring in the human liver where glucuronidation (addition of glucose molecules by the UGT class of enzymes, e.g. UGT1A1) and glutathione addition reactions occur on reactive centres of the xenobiotic.

Phase I and II reactions serve to increase the polarity and reduce the toxicity of the compounds, although many exceptions to the rule are seen. The increased polarity also allows for easy transport of the xenobiotic along aqueous channels.

In the final stage of phytotransformation (Phase III metabolism), a sequestration of the xenobiotic occurs within the plant. The xenobiotics polymerize in a lignin-like manner and develop a complex structure that is sequestered in the plant. This ensures that the xenobiotic is safely stored, and does not affect the functioning of the plant. However, preliminary studies have shown that these plants can be toxic to small animals (such as snails), and, hence, plants involved in phytotransformation may need to be maintained in a closed enclosure.

Hence, the plants reduce toxicity (with exceptions) and sequester the xenobiotics in phytotransformation. Trinitrotoluene phytotransformation has been extensively researched and a transformation pathway has been proposed.

Role of Genetics

Breeding programs and genetic engineering are powerful methods for enhancing natural phytoremediation capabilities, or for introducing new capabilities into plants. Genes for phytoremediation may originate from a micro-organism or may be transferred from one plant to another variety better adapted to the environmental conditions at the cleanup site. For example, genes encoding a nitroreductase from a bacterium were inserted into tobacco and showed faster removal of TNT and enhanced resistance to the toxic effects of TNT. Researchers have also discovered a mechanism in plants that allows them to grow even when the pollution concentration in the soil is lethal for non-treated plants. Some natural, biodegradable compounds, such as exogenous polyamines, allow the plants to tolerate concentrations of pollutants 500 times higher than untreated plants, and to absorb more pollutants.

Hyperaccumulators and Biotic Interactions

A plant is said to be a hyperaccumulator if it can concentrate the pollutants in a minimum percentage which varies according to the pollutant involved (for example: more than 1000 mg/kg of dry weight for nickel, copper, cobalt, chromium or lead; or more than 10,000 mg/kg for zinc or manganese). This capacity for accumulation is due to hypertolerance, or *phytotolerance*: the result of adaptative evolution from the plants to hostile environments through many generations. A number of interactions may be affected by metal hyperaccumulation, including protection, interferences with neighbour plants of different species, mutualism (including mycorrhizae, pollen and seed dispersal), commensalism, and biofilm.

Table of Hyperaccumulators

- Hyperaccumulators table – 1 : Al, Ag, As, Be, Cr, Cu, Mn, Hg, Mo, Naphthalene, Pb, Pd, Pt, Se, Zn

- Hyperaccumulators table – 2 : Nickel

- Hyperaccumulators table – 3 : Radionuclides (Cd, Cs, Co, Pu, Ra, Sr, U), Hydrocarbons, Organic Solvents.

Phytoscreening

As plants are able to translocate and accumulate particular types of contaminants, plants can be used as biosensors of subsurface contamination, thereby allowing investigators to quickly delineate contaminant plumes. Chlorinated solvents, such as trichloroethylene, have been observed in tree trunks at concentrations related to groundwater concentrations. To ease field implementation of phytoscreening, standard methods have been developed to extract a section of the tree trunk for later laboratory analysis, often by using an increment borer. Phytoscreening may lead to more optimized site investigations and reduce contaminated site cleanup costs.

Soil Vapor Extraction

Soil vapor extraction (SVE) is a physical treatment process for in situ remediation of volatile contaminants in vadose zone (unsaturated) soils (EPA, 2012). SVE (also referred to as in situ soil venting or vacuum extraction) is based on mass transfer of contaminant from the solid (sorbed) and liquid (aqueous or non-aqueous) phases into the gas phase, with subsequent collection of the gas phase contamination at extraction wells. Extracted contaminant mass in the gas phase (and any condensed liquid phase) is treated in aboveground systems. In essence, SVE is the vadose zone equivalent of the pump-and-treat technology for groundwater remediation. SVE is particularly amenable to contaminants with higher Henry's Law constants, including various chlorinated solvents and hydrocarbons. SVE is a well-demonstrated, mature remediation technology (Hutzler et al., 1990; Pedersen and Curtis, 1991; Noyes, 1994; Stamnes and Blanchard, 1997; Suthersan, 1999; Kahn et al., 2004; Damera and Bhandari, 2007) and has been identified by the U.S. Environmental Protection Agency (EPA) as presumptive remedy (EPA, 1993, 1996, 2011).

SVE Configuration

The soil vapor extraction remediation technology uses vacuum blowers and extraction wells to induce gas flow through the subsurface, collecting contaminated soil vapor, which is subsequently treated aboveground. SVE systems can rely on gas inflow through natural routes or specific wells may be installed for gas inflow (forced or natural). The vacuum extraction of soil gas induces gas flow across a site, increasing the mass transfer driving force from aqueous (soil moisture), non-aqueous (pure phase), and solid (soil) phase into the gas phase. Air flow across a site is thus a key aspect, but soil moisture and subsurface heterogeneity (i.e., a mixture of low and high permeability materials) can result in less gas flow across some zones. In some situations, such as enhancement of monitored natural attenuation, a passive SVE system that relies on barometric pumping may be employed (Early et al., 2006; Kamath et al., 2010).

Conceptual Diagram of Basic Soil Vapor Extraction (SVE) System for Vadose Zone Remediation

SVE has several advantages as a vadose zone remediation technology. The system can be implemented with standard wells and off-the-shelf equipment (blowers, instrumentation, vapor treatment, etc.). SVE can also be implemented with a minimum of site disturbance, primarily involving well installation and minimal aboveground equipment. Depending on the nature of the contamination and the subsurface geology, SVE has the potential to treat large soil volumes at reasonable costs.

The soil gas (vapor) that is extracted by the SVE system generally requires treatment prior to discharge back into the environment. The aboveground treatment is primarily for a gas stream, although condensation of liquid must be managed (and in some cases may specifically be desired). A variety of treatment techniques are available for aboveground treatment (EPA, 2006) and include thermal destruction (e.g., direct flame thermal oxidation, catalytic oxidizers), adsorption (e.g., granular activated carbon, zeolites, polymers), biofiltration, non-thermal plasma destruction, photolytic/photocatalytic destruction, membrane separation, gas absorption, and vapor condensation. The most commonly applied aboveground treatment technologies are thermal oxidation and granular activated carbon adsorption. The selection of a particular aboveground treatment technology depends on the contaminant, concentrations in the offgas, throughput, and economic considerations.

SVE Effectiveness

The effectiveness of SVE, that is, the rate and degree of mass removal, depends on a number of factors that influence the transfer of contaminant mass into the gas phase. The effectiveness of SVE is a function of the contaminant properties (e.g., Henry's Law constant, vapor pressure, boiling point, adsorption coefficient), temperature in the subsurface, vadose zone soil properties (e.g., soil grain size, soil moisture content, permeability, carbon content), subsurface heterogeneity, and the air flow driving force (applied pressure gradient). As an example, a residual quantity of a highly volatile contaminant (such as trichloroethene) in a homogeneous sand with high permeability and low carbon content (i.e., low/negligible adsorption) will be readily treated with SVE. In contrast, a heterogeneous vadose zone with one or more clay layers containing residual naphthalene would require a longer treatment time and/or SVE enhancements. SVE effectiveness issues include tailing and rebound, which result from contaminated zones with lower air flow (i.e., low permeability zones or zones of high moisture content) and/or lower volatility (or higher adsorption). Recent work at U.S. Department of Energy sites has investigated layering and low permeability zones in the subsurface and how they affect SVE operations [Switzer and Kosson, 2007; Oostrom et al., 2007].

Enhancement of SVE

Enhancements for improving the effectiveness of SVE can include directional drilling, pneumatic and hydraulic fracturing, and thermal enhancement (e.g., hot air or steam injection) (Frank and Barkley, 1995; EPA, 1997; Peng et al., 2013). Directional drilling and fracturing enhancements are generally intended to improve the gas flow through the subsurface, especially in lower permeability zones. Thermal enhancements such as hot air or steam injection increase the subsurface soil temperature, thereby improving the volatility of the contamination. In addition, injection of hot (dry) air can remove soil moisture and thus improve the gas permeability of the soil. Additional thermal technologies (such as electrical resistance heating, six-phase soil heating, radio-frequency heating, or thermal conduction heating) can be applied to the subsurface to heat the soil and volatilize/desorb contaminants, but these are generally viewed as separate technologies (versus a SVE enhancement) that may use vacuum extraction (or other methods) for collecting soil gas.

Design, Optimization, Performance Assessment, and Closure

On selection as a remedy, implementation of SVE involves the following elements: system design, operation, optimization, performance assessment, and closure. Several guidance documents provide information on these implementation aspects. EPA and U.S. Army Corps of Engineers (USACE) guidance documents (EPA, 2001, 2004; USACE, 2002) establish an overall framework for design, operation, optimization, and closure of a SVE system. The Air Force Center for Engineering and the Environment (AFCEE) guidance (AFCEE, 2001) presents actions and considerations for SVE system optimization, but has limited information related to approaches for SVE closure and meeting remediation goals. Guidance from the Pacific Northwest National Laboratory (PNNL) (Truex et al., 2013) supplements these documents by discussing specific actions and decisions related to SVE optimization, transition, and/or closure.

Design and operation of a SVE system is relatively straightforward, with the major uncertainties having to do with subsurface geology/formation characteristics and the location of contamination. As time goes on, it is typical for a SVE system to exhibit a diminishing rate of contaminant extraction due to mass transfer limitations or removal of contaminant mass. Performance assessment is a key aspect to provide input for decisions about whether the system should be optimized, terminated, or transitioned to another technology to replace or augment SVE.

Assessment of rebound and mass flux (Switzer et al., 2004; Brusseau et al., 2010; Truex et al., 2013) provide approaches to evaluate system performance and obtain information on which to base decisions.

Related Technologies

Several technologies are related to soil vapor extraction. As noted above, various soil-heating remediation technologies (e.g., electrical resistive heating, in situ vitrification) require a soil gas collection component, which may take the form of SVE and/or a surface barrier (i.e., hood). Bioventing is a related technology, the goal of which is to introduce additional oxygen (or possibly other reactive gases) into the subsurface to stimulate biological degradation of the contamination. In situ air sparging is a remediation technology for treating contamination in groundwater. Air is injected and "sparged" through the groundwater and then collected via soil vapor extraction wells.

Portable Emissions Measurement System

A portable emissions measurement system (PEMS) is essentially a lightweight 'laboratory' that is used to test and/or assess mobile source emissions (i.e. cars, trucks, buses, construction equipment, generators, trains, cranes, etc.) for the purposes of compliance, regulation, or decision-making. Early examples of vehicle emissions equipment were developed and marketed by Warren Spring Laboratory UK during the early 1990s. This equipment was used to measure on-road emissions as part of the UK Environment Research Programme. Governmental entities like United States Environmental Protection Agency (USEPA), European Union, as well as various states and private sector entities have begun to utilize PEMS in order to reduce both the costs and time involved in making mobile emissions decisions. Various state, federal, and international agencies began referring to this shorthand term in the early 2000s, and the nickname became part of industry parlance.

Background

Since the mid-19th century, dynamometers (or "dyno" for short) have been used to measure torque and rotational speed (measured in rpm) from which power produced by an engine, motor or other rotating prime mover can then be calculated. A chassis dynamometer measures power from the engine through the wheels. The vehicle is parked on rollers which the car then turns and the output is measured. These dynos can be fixed or portable. Because of frictional and mechanical losses in the various drivetrain components, the measured horsepower is generally 15–20 percent less than the brake horsepower measured at the crankshaft or flywheel on an engine dynamometer. Historically though, dynamometer emission tests are very expensive, and have usually involved removing fleet vehicles from service for a long period of time. Also, the data derived from such testing is not representative of "real world" driving conditions, not least due to numerous and ample flexibilities in laboratory test procedures, particularly in the European Union's "New European Driving Dycle (NEDC)". Beyond that, this test method cannot be deemed as representative, especially due to the relatively low amount of repeatable tests at such a facility.

Introduction of PEMS

Leo Breton of the USEPA questioned dynamometer testing's ability to reflected real-world emissions levels of vehicles on-road. To investigate the matter, in 1995 he constructed a device called ROVER, short for Real-time On-road Vehicle Emissions Reporter. The first commercially available fully integrated PEMS device was invented and patented by Michal Vojtisek-Lom, in conjunction with Clean Air Technologies International (CATI) Inc. out of Buffalo, New York in 1999. The first unit was developed for, and sold to - Dr. H. Christopher Frey of North Carolina State University NCSU) for the first ever on-road testing project sponsored by the North Carolina Department of Transportation. David W. Miller, who co-founded CATI, first coined the phrase "Portable Emissions Measurement System" and "PEMS" when working on a New York City Metropolitan Transportation Agency bus project with Dr. Thomas Lanni of the New York State Department of Environmental Conservation in 2000. The phrase PEMS was formally adopted by the USEPA and other governments in 2002.

Many governmental entities (such as the USEPA and the United Nations Framework Convention on Climate Change or UNFCCC) have identified target mobile-source pollutants in various mobile standards as CO_2, NO_x, Particulate Matter (PM), Carbon Monoxide (CO), Hydrocarbons(HC), to ensure that emissions standards are being met. Further, these governing bodies have begun adopting in-use testing program for non-road diesel engines, as well as other types of internal combustion engines, and are requiring the use of PEMS testing. It is important to delineate the various classifications of the latest 'transferable' emissions testing equipment from-time PEMS equipment, in order to best understand the desire of portability in field-testing of emissions.

Defining Portability

An important step in the evaluation of a PEMS device is to define what a PEMS device is as well as to understand various classifications of 'transferable vehicle testing equipment':

Definition of the Term "Portable"

The word portable typically conveys an object that is "Carried or moved with ease, such as a light or small typewriter." The implication is that this is a special attribution for what would otherwise generally be non-portable objects. PEMS is in contrast with lab based systems that measure in circumstances other than normal use.

Definition of the Term "Mobile"

The definition of mobile is essentially "...capable of moving or of being moved readily from place to place: a mobile organism; a mobile missile system."

Definition of "Instrumented"

Instrumented means to be "a device for recording, measuring, or controlling, especially such a device functioning as part of a control system."

Therefore, the subtle difference between 'portable' and 'mobile' is that a portable system is a lightweight device able to be carried, whereas a mobile system can be readily moved, and 'instrumented' means that the testing equipment has been incorporated into the host system. These distinctions are critical, especially considering additional guidelines from various US and International standards.

Definition Determined by the National Institute for Occupational Safety and Health (NIOSH)

The National Institute for Occupational Safety and Health (NIOSH) defines these terms based on an equation known as the "NIOSH Lifting Equation" (http://www.cdc.gov/niosh/docs/94-110/pdfs/94-110-b.pdf) and the "NIOSH Procedures for Analyzing Lifting Jobs (http://www.cdc.gov/niosh/docs/94-110/pdfs/94-110-c.pdf). These clearly outline safety procedures and equipment. (these are also specified in the "Occupational Safety Hazard Act 29 CFR parts 1903, 1904, and 1910)

Safety Guidelines and Standards (the NIOSH Lifting Equation)

It is imperative to refer to existing federal standards and guidelines when determining a proper ergonomically safe and correct procedure. Not only is this important to ensure the safety of the worker(s), but also to ensure the reduction in potential future liability. Therefore, the revised NIOSH Lifting Equation is an excellent source of information to determine what a single worker should or shouldn't perform.

Based upon the NIOSH lifting equation and assuming that this diagram is analogous to the lifting of a PEMS into the cab of a heavy-duty truck the upper threshold of the total weight of a PEMS device typically should not exceed 45 lb (20 kg)., in order to be congruent with national and international safety standards. This not only allows for much more safe maneuverability and ease of use, but it also reduces the amount of workers required to safely perform such tasks.

Economic Advantage of PEMS Equipment

Because a PEMS unit is able to be carried easily by one person from jobsite to jobsite, and can be used without the requirement of 'team lifting', the required emissions testing projects are economically viable. Simply put, more testing can be done more quickly, by less workers, dramatically increasing the amount of testing done in a certain time period. This in turn, significantly reduces the 'cost per test', yet at the same time increases the overall accuracy required in a 'real-world' environment. Because the law of large numbers will create a convergence of results, it means that repeatability, predictability, and accuracy are enhanced, while simultaneously reducing the overall cost of the testing.

On-road Emissions Patterns Identified by PEMS

Nearly all modern engines, when tested new and according to the accepted testing protocols in a laboratory, produce relatively low emissions well within the set standards. As all individual engines of the same series are supposed to be identical, only one or several engines of each series get tested. The tests have shown that:

1. The bulk of the total emissions can come from relatively short high-emissions episodes

2. Emissions characteristics can be different even among otherwise identical engines

3. Emissions outside of the bounds of the laboratory test procedures are often higher than under the operating and ambient conditions comparable to those during laboratory testing

4. Emissions deteriorate significantly over the useful life of the vehicles

5. There are large variances among the deterioration rates, with the high emissions rates often attributable to various mechanical malfunctions

These findings are consistent with published literature, and with the data from a myriad of subsequent studies. They are more applicable to spark-ignition engines and considerably less to diesels, but with the regulation-driven advances in diesel engine technology (comparable to the advances in spark-ignition engines since the 1970s) it can be expected that these findings are likely to be applicable to the new generation diesel engines. Since 2000, multiple entities have utilized PEMS data to measured in-use, on-road emissions on hundreds of diesel engines installed in school buses, transit buses, delivery trucks, plow trucks, over-the-road trucks, pickups, vans, forklifts, excavators, generators, loaders, compressors, locomotives, passenger ferries, and other on-road, off-road and non-road applications. All the previously listed findings were demonstrated; in addition, it was noticed that extended idling of engines can have a significant impact on the emissions during subsequent operation.

Also, PEMS testing identified several engine "anomalies" where fuel-specific NOx emissions were two to three times higher than expected during some modes of operation, suggesting deliberate alterations of the engine control unit (ECU) settings. Such data set can be readily used for developing emissions inventories, as well as to evaluate various improvements in engines, fuels, exhaust after-treatment and other areas. (Data collected on "conventional" fleets then serves as "baseline" data to which various improvements are compared.) This data set can also be examined for compliance with not-to-exceed (NTE) and in-use emissions standards, which are 'US-based' emission standards that require on-road testing.

Accuracy of PEMS

The question often arises as to the target accuracy of PEMS. As PEMS are typically limited in size, weight and power consumption, it is often difficult for PEMS to offer the same accuracy and variety of species measured as is possible with top of the line laboratory instrumentation. For this reason, objections were raised against using PEMS for compliance verification.

On the other hand, fleet emissions deduced from laboratory measurements can be subject to significant inaccuracies if the selected engines and operating conditions were not representative of the fleet, or if deliberate anomalies (i.e., dual mapping of the ECU) were not demonstrated during laboratory testing.

The question of how accurate a monitoring system needs to be therefore cannot be objectively answered, neither can a monitoring system be easily designed, without first considering the intended application of the system and the errors associated with different approaches.

It is expected that a variety of on-board systems will be designed, ranging from suitcase-sized PEMS to instrumented trailers towed behind the tested truck. The benefits of each approach need to be considered in light of other sources of errors associated with emissions monitoring, notably vehicle-to-vehicle differences, and the emissions variability within the vehicle itself. In other words, one needs to consider the total of:

1. The difference between what is measured and what is actually emitted during a test

2. The difference between what is emitted during the test and what the vehicle emits during its everyday duties

3. The difference between the emissions characteristics of the tested vehicle and the overall emissions levels of the entire fleet.

For example, when evaluating a benefit of cleaner fuels on a fleet of city buses, one needs to compare taking a bus out of service, installing a laboratory-grade monitoring system, loading it with sandbags and driving it on a simulated route against testing several buses on their regular routes, with passengers on board, using a simpler (and possibly less accurate) monitoring system.

Additional PEMS Criteria

Another important aspect that needs to be evaluated is the safety of using PEMS on public roadways. Extensions of the tailpipe, lines and cables extending far beyond the vehicle sides, lead-acid batteries located in the passenger compartment of a bus, sharp objects, hot components accessible to bystanders, equipment blocking emergency exits or interfering with the driver, loose components likely to be caught on moving parts, and other potential hazards need to be examined. Also, any modifications to or disassembly of the tested vehicle (i.e., drilling into the exhaust, removing intake air system) need to be examined for their acceptance by both fleet managers and drivers, especially on passenger-carrying vehicles. The source of power for PEMS is a concern, as only a limited amount of power can be extracted from the vehicle electrical system. Sealed lead-acid batteries, fuel cells and generators have been used as external power sources, each with a potential significant hazard when driven on the road. A PEMS also has to be practical. Installation time and the expertise level required to perform installation and to operate the PEMS will have a significant impact on the cost of the test, and on the number of vehicles tested. Versatility (ability to test different vehicles) may be important if testing dissimilar engines or vehicles. Total size, weight and transportability of the PEMS needs to be considered when testing at different locations, including any consumables such as calibration gases. Any restrictions on transport of hazardous materials (i.e.Flame ionization detector (FID) fuel or calibration gases) need to be taken into the account. The ability of the test crew to repair PEMS in the field using locally available resources can also be essential when testing away from the base. Thus, PEMS evaluation protocol should be expanded. In addition to the laboratory comparison testing, which is a measure of how accurately PEMS measures when operated in a laboratory, the accuracy and repeatability of PEMS should also be examined on the road, possibly while driving along a well-defined, repeatable route, or while driving chassis dynamometer cycles on a test track.

PEMS Suitability to Application

Ultimately, it should be demonstrated to show that a PEMS is suitable to the desired application. If the ultimate goal is to verify the compliance with in-use emissions requirements, a fleet of vehicles with known characteristics – including engines with dual-mapping and otherwise non-compliant engines – should be made available for testing. It should be then up to the PEMS manufacturers to practically demonstrate how these non-compliant vehicles can be identified using their system.

Testing Volume and Safe Repeatability

In order to achieve the required amount of 'testing volume' needed to validate real-world testing, three points must be considered:

1. System accuracy

2. Federal and/or state health and safety guidelines and/or standards

3. Economic viability based on the first two points.

Once a particular portable emissions system has been identified and pronounced as accurate, the next step is to ensure that the worker(s) are properly protected from work hazards associated with the task(s) being performed in the use of the testing equipment. For example, typical functions for a worker may be to transport the equipment to the jobsite (i.e. car, truck, train, or plane), carry the equipment to the jobsite, and lift the equipment into position.

Advantages of PEMS

On-road vehicle emissions testing is very different from the laboratory testing, bringing both considerable benefits and challenges: As the testing can take place during the regular operation of the tested vehicles, a large number of vehicles can be tested within a relatively short period of time and at relatively low cost. Engines than cannot be easily tested otherwise (i.e., ferry boat propulsion engines) can be tested. True real-world emissions data can be obtained. The instruments have to be small, lightweight, withstand difficult environment, and must not pose a safety hazard. Emissions data is subject to considerable variances, as real-world conditions are often neither well defined nor repeatable, and significant variances in emissions can exist even among otherwise identical engines. On-road emissions testing therefore requires a different mindset than the traditional approach of testing in the laboratory and using models to predict real-world performance. In the absence of established methods, use of PEMS requires careful, thoughtful, broad approach. This should be considered when designing, evaluating and selecting PEMS for the desired application.

A recent example of PEMS advantages over laboratory testing is the Volkswagen (VW) Scandal of 2015. Under a small grant from the International Council for Clean Transportation Dan Carder and Arvind Thiruvengadam of West Virginia University (WVU) uncovered on-board software "cheats" that VW had installed on some diesel passenger vehicles. The only way the discovery could have been made was by a non-programmed, random, on-road evaluation - utilizing a PEMS device. VW is now liable for over 14 billion USD in fines.

References

- Khopkar, S. M. (2004). Environmental Pollution Monitoring And Control. New Delhi: New Age International. p. 299. ISBN 81-224-1507-5.

- Corcoran, E., C. Nellemann, E. Baker, R. Bos, D. Osborn, H. Savelli (eds) (2010). Sick water? : the central role of wastewater management in sustainable development : a rapid response assessment (PDF). Arendal, Norway: UNEP/GRID-Arendal. ISBN 978-82-7701-075-5.

- Martin V. Melosi (2010). The Sanitary City: Environmental Services in Urban America from Colonial Times to the Present. University of Pittsburgh Press. p. 110. ISBN 9780822973379.

- Colin A. Russell (2003). Edward Frankland: Chemistry, Controversy and Conspiracy in Victorian England. Cambridge University Press. pp. 372–380. ISBN 9780521545815.

- Sharma, Sanjay Kumar; Sanghi, Rashmi (2012). Advances in Water Treatment and Pollution Prevention. Springer. ISBN 9789400742048. Retrieved 2013-02-07.

- Tilley, David F. (2011). Aerobic Wastewater Treatment Processes: History and Development. IWA Publishing. ISBN 9781843395423. Retrieved 2013-02-07.

- Joseph S. Devinny, Marc A. Deshusses and Todd S. Webster (1999). Biofiltration for Air Pollution Control. Lewis Publishers. ISBN 1-56670-289-5.

- Tchobanoglous, G., Burton, F.L., and Stensel, H.D. (2003). Wastewater Engineering (Treatment Disposal Reuse) / Metcalf & Eddy, Inc. (4th ed.). McGraw-Hill Book Company. ISBN 0-07-041878-0.

- Evans, Gareth M.; Furlong, Judith C. (2010-01-01). Phytotechnology and Photosynthesis. John Wiley & Sons, Ltd. pp. 145–174. doi:10.1002/9780470975152.ch7/summary. ISBN 9780470975152.

- Kvesitadze, G.; et al. (2006), Biochemical Mechanisms of Detoxification in Higher Plants, Berlin, Heidelberg: Springer, ISBN 978-3-540-28996-8

- Tchobanoglous, George; Burton, Franklin L.; Stensel, H. David; Metcalf & Eddy, Inc. (2003). Wastewater Engineering: Treatment and Reuse (4th ed.). McGraw-Hill. ISBN 0-07-112250-8.

- "Explaining road transport emissions - A non-technical guide, pages 27 to 37. — European Environment Agency". www.eea.europa.eu. Retrieved 2016-03-01.

- Johson, Dennis (2002-02-13). "ROVER - Real-time On-road Vehicle Emissions Reporter Dennis Johnson, US EPA" (PDF). Real-time On-road Vehicle Emissions Reporter Dennis Johnson, US EPA. US EPA. Retrieved 2016-03-01.

- "Real-time on-road vehicle exhaust gas modular flowmeter and emissions reporting system". patents.google.com. 1999-01-05. Retrieved 2016-03-01.

- Griggs, Mary Beth (20 July 2015). "Mercury Scrubbers On Power Plants Clean Up Other Pollutants, Too". Popular Science. Retrieved 30 July 2015.

- Ashton, John; Ubido, Janet (1991). "The Healthy City and the Ecological Idea" (PDF). Journal of the Society for the Social History of Medicine. 4 (1): 173–181. doi:10.1093/shm/4.1.173. Retrieved 8 July 2013.

- Adler, Tina (July 20, 1996). "Botanical cleanup crews: using plants to tackle polluted water and soil". Science News. Retrieved 2010-09-03.

Permissions

Index

www.ingramcontent.com/pod-product-compliance
Lightning Source LLC
Chambersburg PA
CBHW061301190326
41458CB00011B/3733